First Edition

All Rights Reserved.

Copyright © 2018 William Errol Prowse IV

This book may not be reproduced, transmitted, or stored in whole or in part by any means, including graphic, electronic, or mechanical without the express written consent of the author except in the case of brief quotations embodied in critical articles and reviews.

ISBN-13: 978-1546567110

ISBN-10: 1546567119

Mobile Solar Power

Made Easy!

Be sure to leave an honest review after you read this book!
I would appreciate it greatly. ☺

Check out this book's official website for video tutorials,
product reviews, solar package recommendations and contact information:

http://www.mobile-solarpower.com

Disclaimer

The information in this book is not intended or implied to be a substitute for professional electrical design or installation advice. All content, including text, graphics, images, and information, contained on or available through this book is for general information purposes only. The author makes no representation and assumes no responsibility for the accuracy of the information contained on or available through this book, and such information is subject to change without notice. You are encouraged to confirm any information obtained from or through this book with other sources.

The author does not recommend, endorse or make any representation about the efficacy, appropriateness or suitability of any specific tests, products, services, opinions, professionals or other information that may be contained on or available through this book.

Electricity can be dangerous. Please use common sense and practical safety considerations while working with any electrical system. The author is not responsible or liable for any damages that occur from using the information contained in this book.

Affiliation and Endorsement Disclaimer

Any product names, logos, brands, and other trademarks or images featured or referred to within this book are the property of their respective trademark holders. These trademark holders are not affiliated with the author, this book, or our website. They do not sponsor or endorse this book or any of our online products. The author declares no affiliation, sponsorship, nor any partnerships with any registered trademarks.

Table of Contents

Electricity for Beginners 8

Measuring Electricity 9

Series vs. Parallel 12

Overview of Major Solar Power System Components 13

Solar Power System Design Methods 15

    Fast Method 15

    Lazy Method 16

        The Minimalist 16

        The Classic 400 watt 16

        The Off Grid King 17

        Ultra Lightweight 17

        Low Budget 18

        Dystopian Future 18

Traditional Method 20

    1. Calculating the Load 20

    2. Calculating Battery Bank Size 21

    3. Calculating Solar Array Size 22

        How to calculate the maximum solar array size for a battery 23

        How to calculate the minimum solar array size for a battery 23

        Other solar array sizing tips 24

    4. Calculating Solar Charge Controller Size 25

        Efficiency considerations 26

        Other factors to consider 26

How to Select Solar Power System Components 27

    1. Selecting a Battery 28

    2. Selecting Solar Panels 31

        Flexible Solar Panels 31

    3. Selecting a Solar Charge Controller 33

    4. Selecting an Inverter 35

- 5. Selecting Wire...36
  - 12 Volt Wire Gauge Chart..38
- 6. Battery Bank Voltage Monitors...39
  - How low can you safely discharge your battery?..39
- 7. Fuses and Fuse Holders..40
  - How to Calculate the Fuse Size..41
  - Important locations and ratings for fuses..42
- 8. Other Power Sources...43
  - Shore Charging (plug in chargers)..43
  - Generators..43
  - Wind Turbines...43

How to Install a Solar Power System..44

- 1. How to install a battery bank...44
- 2. How to install a solar charge controller...45
- 3. How to install the solar panels...45
  - Solar Panel Safety Line..46
  - Should you tilt your solar panels?..46
- How to Wire up your Solar Power System..47
  - How to Crimp...49
- 1. Connect all batteries and add the main fuse..52
- 2. Connect solar charge controller to the battery bank..56
- 3. Connect individual solar panels together to create a solar panel array............................58
- 4. Pass solar array wires through the roof and connect them to the solar charge controller...............63
- 5. Connect the inverter and fuse block to the battery bank...65
  - Fuse Block Installation..65
  - Inverter Installation...66
- 6. Battery Monitor Installation..67

Adding DC 12 Volt Appliances..69

- XT-60 connector...71
  - Powering a Laptop without an inverter...72
- Adding efficient interior lights to your vehicle...73
- Switches..74

- Temperature Regulation Appliances 75
  - Cooling your vehicle 75
  - Heating your vehicle 76
  - Other Methods 77
- How to use a Bulk DC-DC Converter 78

Adding AC Appliances 79

Off Grid Internet 80
- 4G LTE Router with High Gain Antenna 80

Smart Home Appliances 81

Solar System Maintenance Schedule 83

Odds and Ends 84
- Efficient Computer Options 84
- Phantom Loads 84
- How to find a phantom load 84
- Storing a Solar Power System 85
- Connecting Different Types of Solar Panels Together 85
- Connecting Different Solar Charge Controllers to One Battery Bank 86
- Solar Electric Cooking and Food Preparation 86
- Solar Water Heating 86
- Should you install a battery isolator? 87
- Increasing solar output by reflecting light onto your solar panels 87

Electricity for Beginners

Electricity is either:

- Produced
- Stored
- Consumed

A solar panel will produce electricity, a battery will store electricity and a set of appliances will consume electricity.

Electricity travels in wires, or conductors, to transfer energy across distance. 2 wires are required to carry electrical force from one location to another.

Electrical energy can be carried in two ways:

- Direct Current (DC Power): The flow of electricity is direct and flows like a river. It comes in one wire and flows out the second wire. One wire is positive, one wire is negative. The differentiation of the positive and negative wire is called the polarity and refers to the electrical charge present in the wires which are used to transmit electrical force.
- Alternating Current (AC Power): Unlike DC, the electrical force in an AC circuit does not flow, but instead vibrates back and forth to carry energy. Imagine how ocean waves can transmit energy over vast distances, without moving the water. Same concept. There is no positive or negative wire in an AC circuit. Instead, the polarity or electrical charge in the wires is constantly alternating.

Alternating current is more efficient than direct current over long distances, but direct current is required if you plan to store the electricity in a battery. Appliances can be designed to use DC or AC power. Some motors and lights can be powered with AC power, but most AC appliances transform the AC electricity into DC electricity, which is much more versatile.

AC electricity is great for sending electrical energy over long distances, and DC electricity is versatile and easy to put to work.

In a solar power system:

- Solar panels produce DC electricity that travels through 2 wires and is stored in a battery.
- DC powered appliances are then connected to the battery with 2 wires so that they will consume the stored electricity in the battery and put it to work.
- If you plan to power AC appliances with a DC battery, you will need to transform DC electricity into AC electricity with a device called an inverter. More on this in a bit.

Measuring Electricity

Electricity is measured with a few metrics:

- **Volts:** Energy Potential, or the size of the force that sends the electricity through the wire. The energy potential, or volts, is always present whether the electricity is being used or not.

    Water Hose Analogy: Volts are similar to the pressure of water in a garden hose. If you hook a spray nozzle to the garden hose and the spray nozzle is closed, the pressure is still present.

- **Amps:** Energy Current, or the amount of electricity going through a wire. The more amps a wire must carry, the thicker the wire must be. Amps are only present when electricity is traveling through a wire or being consumed by an appliance.

    Water Hose Analogy: Current is similar to the rate of flow. Think of it as the total amount of water that a hose can carry. If a hose is thicker, it can carry more water. The rate of flow can only occur when water is moving through the hose.

- **Watts:** The total "power" produced. This is the measurement that combines Volts and Amps.

    Water Hose Analogy: How fast you can fill a bucket with water.

The components of a solar power system will produce electricity, store electricity or consume electricity. We can use volts, amps, and watts to describe how much electricity something produces, stores or consumes.

How to use the volt/amp/watt rating in a solar power system:

- **The voltage rating will determine the compatibility of a component.** If a battery is rated for 12 volts, it can only power 12-volt appliances. There are exceptions to this, but to keep things simple, use the voltage to determine whether one component will work with another component.
- **The amp rating will determine how much electricity is produced/stored/consumed at a given voltage.** In a solar system, we will use the amp rating of a component to determine what thickness of wire is required to attach it to the system. The more electricity a component produces or consumes, the thicker the wire has to be to connect it to the system. Some components will have a voltage rating and an amp rating. More on this later.
- **The watt rating will be used to figure out the total amount of electricity a component is producing/storing/consuming at a given moment.**

When electricity is being produced or consumed, the volt and amp rating will determine the watt rating. You can figure out how many watts a system component is producing or consuming by multiplying its amp rating by its voltage rating.

Amps x Volts = Watts

- A solar panel that is producing 5 amps of electricity at 20 volts will produce 100 watts
- A solar panel that is producing 2 amps of electricity at 40 volts will produce 80 watts

- A small fan consuming 10 amps of electricity at 12 volts, will be consuming 120 watts
- A small fan consuming 5 amps of electricity at 12 volts, will be consuming 60 watts

Calculating the wattage of a component is useful, but we need to take it one step further. If you combine the watt rating of a component, and how long you use it for, you will determine the components "watt hour rating".

Wattage x Hours being used = Watt hour rating

Multiply the wattage of a component by how many hours it is being used for:

- A 100 watt solar panel producing power for 3 hours will produce 300 watt hours
- A 500 watt solar panel array producing power for 1 hour will produce 500 watt hours
- A 1000 watt solar panel array producing power for 30 minutes will produce 500 watt hours

- A 1000 watt microwave being used for 1 hour will use 1000 watt hours
- A 100 watt fan that is used for 10 hours will use 1000 watt hours
- A 1000 watt microwave being used for 15 minutes will use 250 watt hours

If you are determining the watt hour capacity of a battery, or how much electricity a battery can store, you must determine the watt hour rating manually.

Batteries available for purchase are typically rated in "amp hours". This figure represents how many amps can be used in one hour, at a specified voltage rating.

If a battery is rated for 200 amp hours at 12 volts, the battery can store enough electricity to produce 200 amps of electricity, at 12 volts, for 1 hour.

If you increase the duration of consumption, you decrease the amps that the battery can supply:

- A 12 volt, 200 amp hour battery can produce 100 amps of electricity for 2 hours
- A 12 volt, 200 amp hour battery can produce 50 amps of electricity for 4 hours
- A 12 volt, 200 amp hour battery can produce 25 amps of electricity for 8 hours

The watt hour rating of a battery is the amp hour rating multiplied by the voltage of the battery:

- A 12 volt, 200 amp hour battery (12 volts x 200 amp hours = 2400 watt hours) can power 2400 watts for 1 hour. This battery has a watt hour rating of 2400 watt hours.
- A 12 volt, 50 amp hour battery (12 volts x 50 amp hours = 600 watt hours) can power 600 watts for 1 hour. This battery has a watt hour rating of 600 watt hours.

Now that we can calculate the wattage and watt hour rating, a solar power system will be easier to design. Here is a full system watt usage example:

- A solar power system has 4 solar panels that produce 5 amps each, at 20 volts, which means each solar panel produces 100 watts (5 amps x 20 volts = 100 watts). There are 4 solar panels, so the total power that this solar array can produce at a given moment is 400 watts.
- When exposed to full sunshine for 4 hours, the 400 watt solar panel array will produce 1600 watt hours (400 watts x 4 hours = 1600 watt hours)
- The battery bank that the solar panels need to charge is rated for 133 amp hours at 12 volts. This means that the battery bank can store 1596 watt hours (133 amp hours x 12 volts = 1596 watt hours). This means that in 4 hours of full sunshine, the solar panels can charge this battery.
- If we use the battery bank to power a 70 watt fan, we will be able to use it for 22.8 hours. (1596 watt hour battery rating divided by 70 watts = 22.8) If instead, we use the battery bank to power a 1500 watt microwave, we will be able to power it for around 1 hour.

Now that you understand the basics, we can use these formulas to determine how much electricity we will need and how large our solar power system components should be.

Formulas Summarized:

- Amps x Volts = Watts
- Watts x Hours being used = Watt hour rating
- For batteries: Amp hour rating x Voltage of the battery = Watt hour rating

---

Series vs. Parallel

Now that we understand volts and amps, we can go over different wiring configurations:

In Series: Daisy chain configuration. Attach the negative lead of one component to the positive lead of another component.

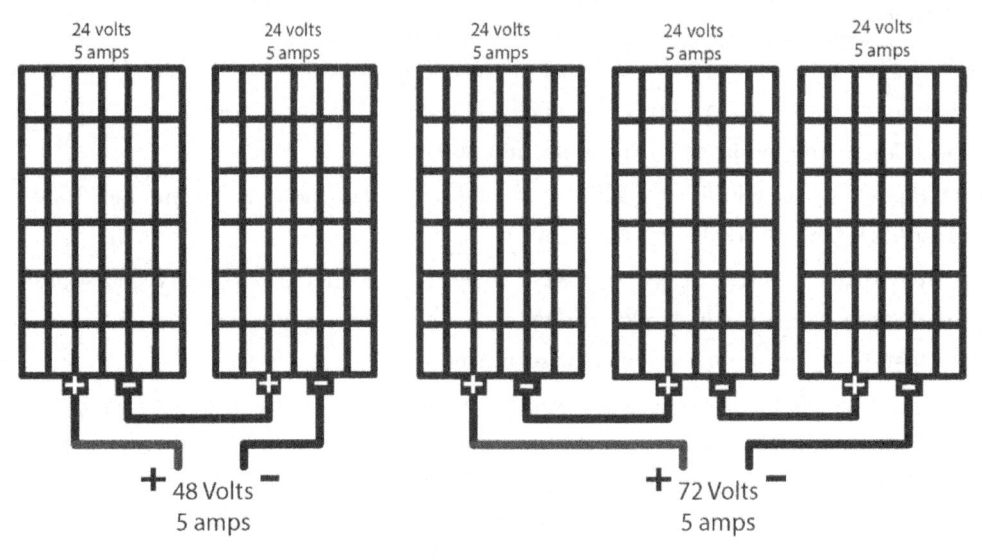

In a series configuration, the voltage increases, but the amps do not change.

In Parallel: Attach the negative lead of one component to a negative lead of another component, and the positive lead to the positive lead of another component.

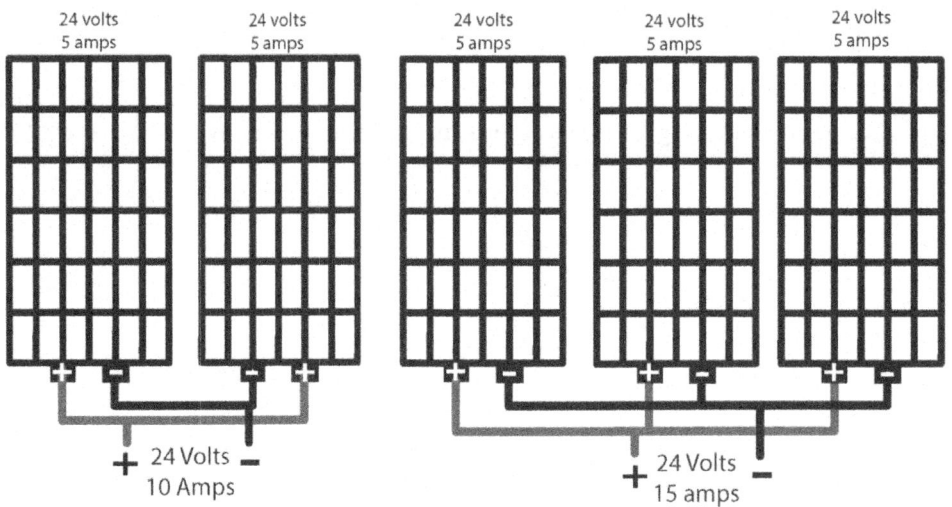

In a parallel configuration, the voltage does not change, but the amps increase.

No need to memorize these configurations right now. Just realize that you can change the voltage and amp rating of a system component by wiring it in different ways. More on this later.

Overview of Major Solar Power System Components

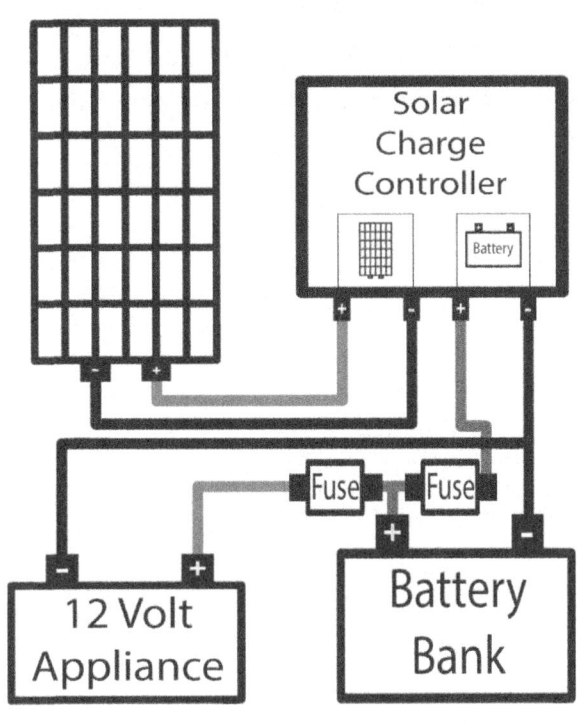

Solar Panel: Creates electricity with sunshine. Sturdy construction enables solar panels to withstand heat, pressure, rain, snow and more. A typical 12-volt solar panel will produce 17-24 volts (typically 20 volts).

Battery Bank: Stores electricity created by solar panels. Most battery banks for vehicle applications are going to be 12 volts. You can use 6/24/48 volt battery banks, but 12 volt is most common.

Solar Charge Controller: Charges the battery bank with power created by the solar panels. Solar panel electricity is not usable for most applications because it varies constantly. The solar charge controller takes the constantly changing 0-24 volts produced by the solar panels and produces a constant voltage that is suitable to charge a 12-volt battery bank, which is typically 12.6-14.5 volts.

There are two types of solar charge controllers: PWM (pulse width modulation) and MPPT (maximum power point tracking). We will cover these details later.

12 volt Appliances: What will ultimately consume the electricity that your system creates. These can be LED lights, USB chargers, stereo, microwaves, computers, backup monitors, seat heaters and more!

Wires: Connects everything together. An important and often overlooked component of a solar power system.

Other Components:

Inverter: Converts 12 volt DC power to 110 volt AC power. This enables you to plug traditional AC appliances into your DC solar power system.

Battery Monitor: Shows the voltage of your system which can allow you to estimate current battery bank capacity, load draw (how much electricity an appliance is using) and more.

Alternative Energy Sources: There are other ways to produce 12-volt electricity. The most common are generators, battery isolators (using your vehicle's engine alternator as a generator), wind turbines and shore power plug-in battery chargers.

Solar Power System Design Methods

1. **Fast method:** Follow "rules of thumb" to determine the size of your system
2. **Lazy method:** Choose one of my pre-calculated systems
3. **Traditional method:** Do a little math and manually calculate the size of your system (recommended)

Smart method: Skim through the "fast method" and "lazy method" to get a feel for everything, and then use the "traditional method" to calculate your own personal system.

Note: If you use the "Fast Method" or "Lazy Method", you will still need to manually calculate the proper gauge of wire and fuse sizes for your system.

Fast Method

- **Fill the roof with as many solar panels as possible:**
  Vehicles have a limited roof space for solar panels. Build your system around this constraining factor
- **Buy a solar charge controller that can handle the power produced by your solar panels:**
  -If you have limited roof space (100-250 watts of solar), and do not plan to add more panels in the future, use a 20 amp MPPT solar charge controller
  -For most mobile systems (300-450 watts of solar), use a 40 amp MPPT solar charge controller
  -For large systems (450-700 watts of solar), use a 60 amp MPPT solar charge controller
  - For extra-large systems (700-950 watts of solar), use a 80 amp MPPT controller
  (80 amp controllers cost a lot, so it is usually cheaper to buy 2x 40 amp controllers)
- **Determine the size of the battery bank:**
  -For every 100 watts of solar panels you have on your roof, you should have around 75-100 amp hours of sealed lead acid battery. If you are using a lithium battery, buy the largest size you can afford.
- **Find the "safe charging rate" of your battery bank in amps, and make sure that it is more than the total amps produced by your solar panel array.** You can calculate this manually, or you can read the battery manual. Most solar specific battery manuals and data sheets will tell you how many solar panels you can safely connect to a battery. If it is not listed, you can call the manufacturer and ask them. Be sure to check this metric before you buy your battery bank. Each battery has a different charge rate. If you plan to build a large system and need a high charge rate, put multiple smaller batteries together in parallel. This will increase the charge rate more than buying a larger battery.
- **Buy the largest inverter that you can afford:**
  -If you plan to only power a laptop and fan, buy a 750 watt inverter
  -Most people do well with a 1500 watt inverter
  -If you want to run most household appliances, buy a 2000 watt inverter
  -If you want to run multiple large appliances, buy a 3000+ watt inverter

Lazy Method

Want to design your system without using your brain? You have 6 pre-calculated options to choose from:

- **The Minimalist**
  (Minivan or Car)
- **The Classic 400 watt**
  (Large Van or RV)
- **The Off Grid King**
  (Large RV or Bus)
- **Ultra lightweight**
  (Backpacker or Cyclist)
- **Low Budget**
- **Dystopian Future**

The Minimalist

This setup works well if you:

- Plan to install in a minivan/car
- Want a lightweight setup
- Plan to run a small laptop, USB charger, fan, LED lights and other small appliances

Recommended components:

- 12 volt, 100-150 amp hour deep cycle battery
- 200 watts of solar panels and a 20 amp MPPT charge controller
- 750 watt inverter

The Classic 400 watt

This setup works well if you:

- Plan to install in a large van or normal sized RV
- Plan to run a TV, fridge, large laptop and more, every single day

Recommended components:

- 400 watts of solar panels and a 40 amp MPPT charge controller
- 12 volt, 300 amp hour battery bank (keep in mind that this battery bank will weigh around 190 pounds) or a single 100-200 amp hour lithium battery (this battery will weigh 30-60 pounds)
- 1500 watt inverter

The Classic 400 watt is my favorite size. Most people will love this setup!

The Off Grid King

This setup works well if you:

- Plan to use a Large RV or Bus
- Want to power large appliances, such as power tools and microwaves for prolonged durations
- Weight and aerodynamics is not a huge factor for your vehicle

Recommended components:

- 800 watts of solar panels and a 80 amp MPPT solar charge controller
  (or 2x 40 amp MPPT solar charge controllers)
- 12 volt, 350-500 amp hour lead acid battery, or 300 amp hour lithium battery (Choose the largest battery that you can afford to buy and can safely carry in your vehicle. These batteries are heavy and expensive!)
- 2000 watt inverter, or larger

Ultra Lightweight

This setup works well if you:

- Are backpacking and need to travel in cars, trains, bicycles and planes
- Live in a small vehicle and are unable to carry a heavy battery bank
- Cannot permanently mount a solar panel array
- Need to charge your cellphone, a laptop, camera and other travel tools

Recommended components:

- A fold-up solar panel that produces 20-45 watts, with a USB cable output
- USB Battery Banks (2-4) with fast charge capabilities (4 amps input)
- Fast Charge USB AC adapter (to charge the battery banks when there is no sunlight)
- Laptop/camera/phone that can charge with a USB. If you have a powerful laptop, use a portable laptop external battery charger
- Available online are USB powered lights, fans, hand warmers and more

Note: Use the fold up USB solar panel or AC USB adapter to charge the USB battery banks. This entire setup can easily fit in a backpack and can be used in vehicles, or while backpacking.

Low Budget

This setup works well if you:

- Are broke and want a solar system now!
- Willing to sacrifice efficiency and longevity

Recommended components:

- Buy as many solar panels as you can afford. A Chinese manufactured 100 watt solar panel can be acquired for $100 USD.
- Buy a PWM (pulse width modulation) solar charge controller. A 30 amp PWM controller can be purchased for 15 dollars online.
- Buy refurbished forklift batteries. This is how I acquired 2x 225 amp hour, 6-volt forklift batteries for 80 dollars. Online search "refurbished batteries in San Jose" or wherever you live.
- Buy car jumper cables from a thrift shop to wire it all up. Make sure that the jumper cables are 8 gauge to be on the safe side.
- Buy a cheap inverter online. Some 750 watt inverters can be purchased for 40 bucks!

Notes: this works surprisingly well. It will probably not last for longer than 2 years, but it will work.

Dystopian Future

This setup works well if you:

- Are fighting for your life and require electricity to increase your chances of survival

Recommended components:

- Find any and all 12-volt batteries (avoid ones with visible damage) and wire them in parallel. This is usually not smart to do (wiring batteries of different types/ages), but if you are desperate, you are going to do it. Store these batteries in a secluded location, so that if they do explode, they will be far from danger. Try to find a dry location, such as a lifted concrete slab with a roof of some kind.
- Find any and all solar panels. You can salvage them from telephone boxes, parking meters, and roadside construction lights.
- Find a 12-volt regulator of some kind to be used as a solar charge controller. Car alternator voltage regulators are not ideal but can work. DC-DC converters can do the job. You can use a microcontroller and some transistors to craft your own, but that is a bit involved. If you are able to salvage solar panels from a telephone box, use the regulator it comes with.
- Use alternators to build 12-volt hydroelectric dams, wind turbines and more.
- Wire the batteries with whatever kind of conductor you can find. If you find sheet metal, you can cut it and bolt it to the battery terminals.

This is all speculation ☺ I hope none of you will need this setup! It is a good thought exercise.

Before we dive in, look at the schematic below a few times. This is a blueprint for "The Classic 400 watt" system. This is the most common size I have seen on RV's, and people love it!

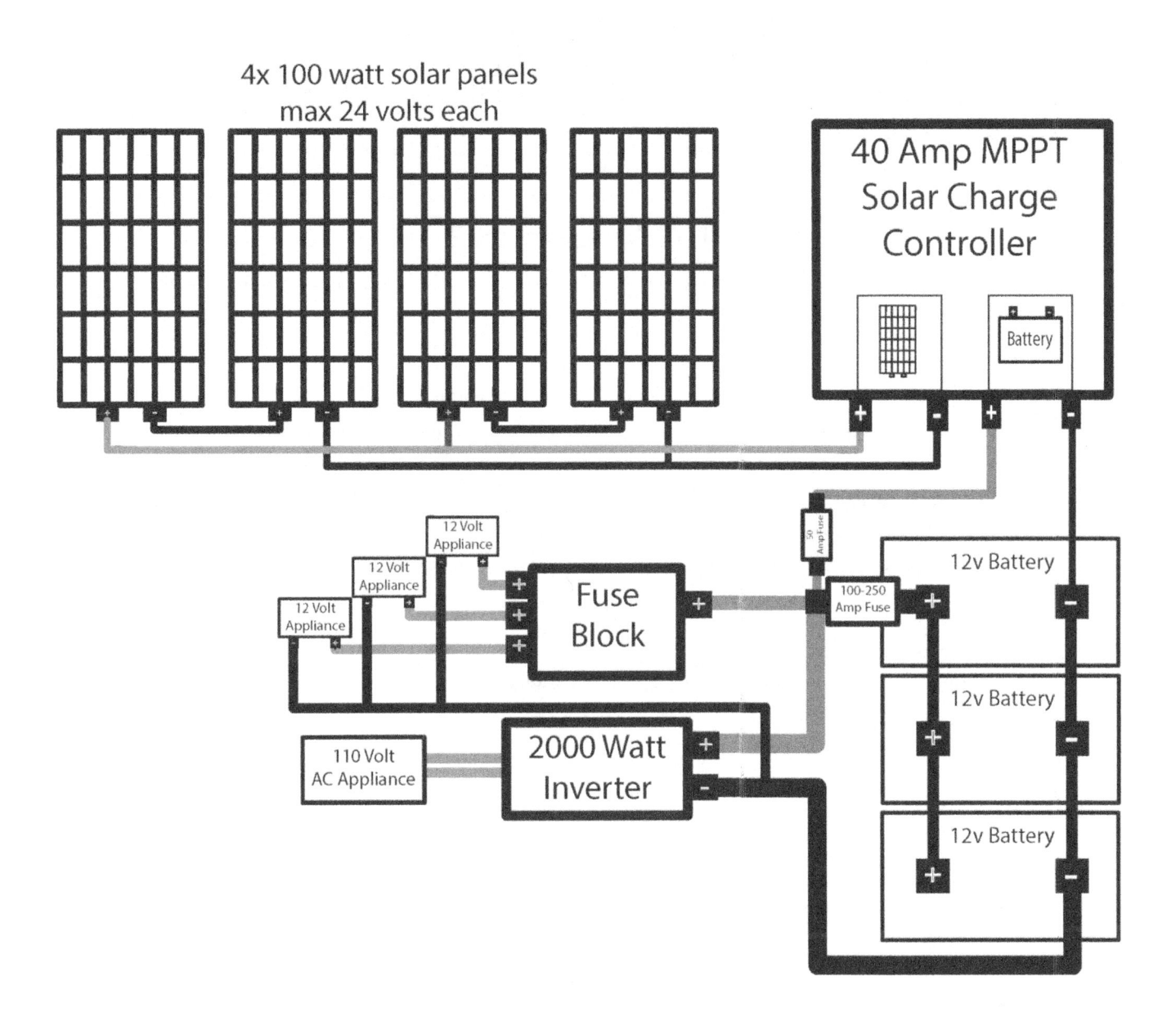

Traditional Method

1. Estimate the daily load (how much electricity your appliances will consume in 1 day)
2. Use the estimated daily load to calculate the battery bank size
3. Use the battery bank size to calculate how many solar panels you need
4. Use the solar panel array size to calculate the solar charge controller size

1. Calculating the Load

To determine how much electricity you need, you will need to figure out the total watt hour requirement of all the appliances you plan to run.

Examples:

- If you use a 100 watt light bulb for 1 hour, it has used 100 watt hours.
- If you have a 30 watt light bulb, and you run it for 3 hours, it will use 90 watt hours.
- If you run a 1000 watt microwave for 30 minutes, it will consume 500 watt hours.

Most appliances will tell you how many watts they use. If you are unsure, look for a label on the back or bottom of the appliance, usually located where the cord attaches to the appliance. It will tell you the voltage, amperage, and wattage. If the appliance only shows the volts and amps, you can determine the wattage by multiplying the volts and the amps:

Amps x Volts = Watts

- A phone charger outputs 5 volts at 2 amps, so it uses 10 watts.
- A LED light strip uses 12 volts at 5 amps, so it uses 60 watts.
- An AC microwave uses 110 volts at 10 amps, so it uses 1100 watts.

Find the wattage of the appliances that you plan to use every day. Estimate the amount of time that you plan to use the appliances, and calculate their watt hour rating:

Watts x Hours used = Watt hour rating

- 60 watt LED light strip used for 4 hours a day= 240 watt hours
- 70 watt fridge that is powered on for 24 hours, but runs for 4 hours (compressor activation is intermittent)= 280 watt hours
- 60 watt laptop used for 6 hours= 360 watt hours
- 1000 watt microwave used for 15 minutes= 250 watt hours

Take all of the watt hour estimates for your appliances and add them together. For this example, we will add the watt hour estimates together from the list above:

Total Appliance Load for One Day: 1130 watt hours

2. Calculating Battery Bank Size

Now that we know how much power we need daily, we can calculate the size of the battery bank.

First, round our daily power requirement of 1130 watt hours to 1200 watt hours (to make the math easier).

Now we need to estimate how much backup power we want to have. Winter time, rainy days, and shady parking spots will reduce the power produced by your solar panels. The battery bank should be large enough to compensate for these times.

Idealistically, you want as large of a battery bank as possible. But because this is a mobile system, and weight is a factor, I would recommend 2 days of power as a backup. If you need a battery bank for a cabin or other stationary structure, 2-5 days of backup power is typical.

Daily appliance load of 1200 watt hours x 2 (days of backup) =
2400 watt hour battery bank required

But here's the catch! Lead acid batteries (most common type of mobile solar battery) can be safely discharged to only 50% capacity without causing damage. A lithium battery, which we will talk about later, does not have this problem.

So if you require a battery that can deliver 2400 watt hours of power, you will need either:

- A 4800 watt hour lead acid battery
- Or a 2400 watt hour lithium battery

Later we will discuss the differences between these two batteries. For now, we just need to calculate the size of the battery bank.

3. Calculating Solar Array Size

Because space is limited on the roof of most vehicles, filling your roof with as many solar panels as you can safely fit is usually the best option.

But keep in mind that if you have too many solar panels, you may accidentally charge your battery bank too fast, which will reduce the life of the battery bank. So if you fill your roof with solar panels, make sure that you build a battery bank large enough to handle it. This applies mainly to lead acid batteries (which have a lower charge rate when compared to most lithium batteries).

Lead acid batteries also need to be fully charged after every use. They also like it when they are fully charged once a day. To keep your batteries healthy, you need a solar panel array that is large enough to charge your lead acid battery bank in one day (in six hours of full sunshine).

A lithium battery does not require a daily charge after it is used, and only needs a full charge every couple of months.

Solar Array Estimates (no math required):

For a single 1200 watt hour lead acid battery (which is a 100 amp hour, 12 volt AGM sealed battery with a max charge rate of 35 amps at 12 volts), use:

- A minimum of 200 watts of solar panels
- A maximum of 400 watts of solar panels

For a single 1200 watt hour lithium battery (which is a 100 amp hour, 12 volt lithium iron phosphate battery with a max charge rate of 100 amps at 12 volts), use:

- No minimum solar array size. Just be sure to fully charge it every couple of months
- A maximum of 1200 watts of solar panels

The suggestions above are just estimates! Each battery bank will have a slightly different charge rate. Be sure to check your batteries manual to see what it recommends. Most solar application batteries will give you a minimum and maximum solar array size recommendation.

You have probably realized that lithium batteries work well with nearly any size of mobile solar panel array. This is usually true, but be sure to check the manual. The charge rate of a lithium battery is dependent on how the battery is designed. Most can handle large charge rates, but not always.

This is not the case with deep cycle lead acid batteries. They usually have consistent charge rates.

But the charge rate of lead acid batteries can change depending on how many you are using. If you parallel connect multiple small lead acid batteries, the charge rate will usually be much higher than if you were to use a single, large lead acid battery (unless the large battery is designed to handle a fast charge rate. But typically, having smaller batteries in parallel will be faster).

The estimates above will give you a general idea of your solar array size. Ultimately, the individual battery charge rate will determine how many solar panels you can attach to it. If you are lazy or smart, call the battery manufacturer and ask them how many solar panels they recommend.

How to calculate the maximum solar array size for a battery

If your batteries manual does not list how many solar panels you can safely use with it, or you want to calculate it manually, we can figure it out. You will need to read the batteries manual (or data sheet that can be found online) and find the "maximum safe charging rate" in amps. As long as the maximum power produced by the solar panels is less than the maximum charge rate of the battery bank, we will be good to go.

Maximum Solar Power < Maximum Charge Rate of Battery Bank

In order to find the maximum power produced by a solar array, we divide the total solar panel watt rating by the voltage of the battery bank.

Example:

- If we have 400 watts of solar panels in a system, divide this number by the voltage of the battery it plans to charge, which is typically 12 volts
- 400 watts divided by 12 volts = 33.3 amps

33.3 amps is the maximum amount of current that our 400 watt solar power system can produce at 12 volts. A typical 100 amp hour, 12 volt lead acid usually should be able to handle 35 amp charge rate.

35 amps is larger than 33.3 amps, so we are good to go!

If you plan to wire multiple 12 volt lead acid batteries together in parallel, you can add the maximum charge rates together. Let's say you have 3 batteries that can each handle 35 amps each. If you wire them in parallel, they can handle a combined maximum charge rate of 105 amps!

How to calculate the minimum solar array size for a battery

For this calculation, we need to know how much solar power is required to charge the battery bank in 6 hours of full sunshine. This will allow the battery bank to charge to full capacity every day.

Divide the usable watt hours of your battery bank by 6:

Battery bank size in watt hours / 6 = Minimum solar array size

- Your battery bank has a total usable capacity of 1200 watt hours. Dividing this number by 6 will give you 200. So for this battery bank, the solar array should be at least 200 watts in size.

- Your battery bank has a total usable capacity of 2000 watt hours. Dividing this number by 6 will give you 333. So for this battery bank, the solar array should be at least 333 watts in size.

If you are using lead acid batteries, determining the minimum solar array size is important because lead acid batteries require a full daily charge cycle to prolong the life of the battery. If your solar array cannot charge your battery bank within 6 hours, you risk a reduction in lead acid battery bank life. If you have a lithium battery, this factor is not important.

Other solar array sizing tips

We need to consider the real world output of a solar panel. Many solar panels that are rated for 100 watts usually produce about 70 watts in full sunshine. We still need to calculate for a system that has 100 watt solar panels, so that the system can handle the power if it is ever produced.

If you are strapped for cash, it is ok to start with the minimum solar array size and build your way up. If I was shooting for a 600 watt solar array, but I could not afford it yet, I would install a 400 watt solar array first. You may find that a 400 watt solar array is plenty for your needs! Just be sure to buy a larger than needed solar charge controller so that you can always add more solar panels or batteries when necessary.

Solar power output is largely determined by where you live. If you live close to the equator, you will obviously have more power. The angle of the panels, time of day and weather conditions will also determine how much power your solar array will produce.

If you live far from the equator, your solar panels may never create the power they are rated to produce, so you may need to experiment with "over-paneling" your system. Over-paneling allows you to wire 2 to 3 times the amount of solar panels to your system, without damaging the charge controller. This requires using a solar charge controller that has this capability, or using a fuse between the solar array and the solar charge controller.

Example:

You live in Alaska and your 100 watt solar panels only produce 40 watts in full sunshine. So instead of using a 400 watt solar panel array, you decide to use an 800 watt array and a solar charge controller that has over-paneling protection. This will enable you to harvest more power from the sunshine available to you.

If you cannot find a solar charge controller that has over-paneling protection, use a fuse to protect the charge controller. If you have a 40 amp MPPT controller, and you wish to over-panel it with 800 watts of solar panels, you will need to calculate the fuse size for the voltage that your panels produce. This is for advanced users only! If the fuse is not the correct size, you will destroy your solar charge controller.

4. Calculating Solar Charge Controller Size

There are 2 variables that will determine the size of your controller:

1. **The solar power array size will determine the "amp rating" of the solar charge controller.** Solar charge controllers are rated in "amps" and this rating refers to how much current (in amps) the controller can create at your battery bank's voltage. The more solar panels you have in your system, the larger the controller needs to be. If you buy a 40 amp charge controller, the maximum charge it can deliver at 12 volts, is 40 amps. The amp rating does not refer to the amp rating of your solar panels.

   To calculate the amp rating of your controller, take the total solar panel array wattage and divide it by the voltage of your battery bank. This will give you the minimum amp rating of your controller.

   Solar Panel Array Wattage / Battery Bank Voltage = Minimum Solar Controller Amp Rating

   Example:
   Your solar array is 400 watts and your battery bank voltage is 12 volts.
   <u>400</u> (**watts of solar on your roof**) / <u>12</u> (**voltage of your battery bank**) = <u>33.3 amps</u> (**minimum amp rating of your solar charge controller**)

   Controllers are usually sold in amp rating increments of 10 and 20. If you go online, it is easy to find controllers that are rated for 10/20/40/60/80 amps. If we need to find a controller that can handle at least 33.3 amps, we should use a 40 amp controller. It is usually a good idea to buy a larger than necessary controller, just in case you wish to add more solar panels in the future.

2. **The maximum input voltage rating of the controller.** If your solar panel array creates a voltage that is larger than the controller can handle, the controller will be damaged. Usually, you do not need to worry about this figure unless your system is very large, or you are wiring panels in series and producing hundreds of volts. For most mobile systems, the maximum rated voltage will not be exceeded (you should still check the manual of your solar charge controller to be on the safe side).

   Typical controller input voltage ratings are 70-150 volts (but be sure to check your manual).

To summarize:

- For small systems (100-250 watts of solar), use a 20 amp controller
- For most systems (300-450 watts of solar), use a 40 amp controller
- For large systems (450-700 watts of solar), use a 60 amp controller
- For extra-large systems (700-950 watts of solar), use a single 80 amp controller
  (80 amp controllers cost a lot, so it is usually cheaper to buy 2x 40 amp controllers)

Efficiency considerations

The math given earlier is great for estimating a battery bank and solar array size, but it will not tell you the true output of your system. Without going into too many details, consider that:

- On average, you will have a 2%-5% wire loss (they give off a small amount of heat)
- Solar Charge Controllers produce heat and create a 2%-30% loss
- Storing electricity in a battery will experience a 1%-15% loss (unless the battery is damaged or old, then it will be more)
- When you use an inverter, you will have a 10%-15% loss (sometimes larger)
- Appliances are not entirely efficient, and they use various regulators and resistors that give off heat. Expect another 1-5% loss.
- Solar panel efficiency drops if they are too hot. This can vary depending on the panel and how it is mounted, and materials used to make it, but it's another efficiency factor to keep in mind.
- One bad connector will choke an entire solar system. The losses can be huge! All connectors, which connect the wires to the batteries/charge controller, need to be crafted properly. To check them, feel them with your hands to see if they are getting warm. All connectors and wires should be cold to the touch (unless they are carrying a lot of electricity, such as during full sunshine or during inverter operation).

So what I like to say is that if you have a 100 watt solar panel on your roof, you only have 50 watts of usable power. This only applies if you have a properly designed system. If you use cheap parts, small wires, or have bad connections, you will not have much power at all. I would not be surprised to see a 100 watt panel producing only 20 watts on a badly designed system.

No matter how perfect your math is in planning your system, you will always have losses and you will need to create a system that is slightly larger than what you need.

When you design a system, do it right from the beginning and you will save yourself from months of frustration and problems. A properly designed system is also safer and the chance of experiencing an electrical fire is practically non-existent.

Once the system is installed, you don't have to think about it! You have free electricity for years, and it's amazing. I am currently writing this book with solar power, and it's awesome!

Other factors to consider

Adding solar panels to any vehicle causes changes in the aerodynamic profile, which can change the efficiency (miles per gallon) of your vehicle. If you plan to travel constantly, you will need a slightly different system than someone who stays stationary. Also remember that system components can be heavy, especially the battery bank. The heavier your vehicle is from carrying a large battery, the harder it is for your vehicle to stop, and the harder the engine and transmission has to work to move the vehicle.

After you build your system, you may need more power. This is very easy to do with proper planning. You simply add more solar panels and/or batteries. Try to create the system in a way so that it is easy to expand it, such as buying a larger solar charge controller than needed, or using larger gauge wires than necessary. This will ensure that your system is scalable to some degree, or working to its full potential.

How to Select Solar Power System Components

Now is the time to go online and order everything you need. After you have calculated your system, you probably have created a small shopping list of components and their ideal size:

- 3600 watt hour battery bank
- 400 watts of solar panels
- 40 amp solar charge controller
- 2000 watt inverter

The components above require some calculation when choosing, but there are a few more that we must now add to the list:

- Wires
- Fuse blocks and fuses
- Battery monitor
- Tools
- Other power sources

Now we will go over the key factors to consider when selecting each of these components. Each component requires some time and research to select properly so that it will work with your system.

After you choose your battery bank, solar panels, charge controller, and inverter, you can then calculate the type and thickness of wire and the type and size of fuse to use with your system.

If you want to avoid the headache of shopping for these components online, check out my website where I recommend the latest and greatest solar products. It can be found at: http://www.mobile-solarpower.com

1. Selecting a Battery

The world of battery chemistry is vast, and several books could be written on this topic alone. To avoid confusion, we will not dive too deep into this topic. What you need to know is that most batteries are not fit for solar application. Solar battery banks require:

- Long lifespan (7-25 years)
- Large Depth of Discharge Rating (how much usable electricity is available in a battery)
- High cycle life (how many times you can discharge and recharge the battery)

So what we need is a deep cycle battery. These are made to have a large capacity, high cycle life, and long life span. There are many different kinds of deep cycle batteries, but for most solar applications, you should use an AGM sealed deep cycle battery or a lithium battery (lithium iron phosphate chemistry). These are the safest, most efficient, and highest capacity batteries on the market, and are designed for solar applications.

Many people make the mistake of using a car or marine battery as a solar battery. These batteries are not designed for solar and fail miserably. These batteries have a very low depth of discharge rating. This means that you can only run these batteries down to 95% of capacity, and no lower. They are designed to create a lot of electricity in a short amount of time which is able to start an internal combustion engine. They are not designed to power a load for a prolonged duration.

To give you an idea:

- A 100 amp hour car battery only has 5 amp hours of usable capacity
  (total capacity of the battery is <u>5 amp hours</u>)
- A 100 amp hour AGM sealed deep cycle lead acid battery has 50 amp hours of usable capacity
  (total capacity of battery is <u>50 amp hours</u>)
- A 100 amp hour Lithium deep cycle battery has 100 amp hours of usable capacity
  (total capacity of battery is <u>100 amp hours</u>)

Lithium batteries have the best depth of discharge around. They can be safely discharged to 0% (if the battery management system allows for it). If you want to increase the life of a lithium battery, discharge it to 20% instead.

So you have two options:

- AGM sealed deep cycle battery
- Lithium battery (specifically lithium iron phosphate)

The king of solar batteries is lithium, hands down. But you may not be able to afford it. AGM batteries work well and are cheaper up front.

Lead Acid vs. Lithium Battery Comparison

Keep in mind that a 200 amp hour AGM sealed battery has the equivalent usable capacity of a 100 amp hour lithium battery.

AGM Sealed Battery Price Estimate and Weight:

- 100 amp hour battery is around 170 USD and weighs 60 pounds
- 155 amp hour battery is around 290 USD and weighs 90 pounds
- 200 amp hour battery is around 400 USD and weighs 120 pounds

Lithium Battery Price Estimate and Weight:

- 100 amp hour battery is around 900 USD and weighs 35 pounds
- 200 amp hour battery is around 1800 USD and weighs 62 pounds

So lithium seems to cost a lot for the capacity, initially. But Lithium is surprisingly cheaper because it has an increased rated life cycle compared to AGM sealed batteries:

- If you discharge an AGM sealed battery to 50%, you will get around 500 charge cycles
- If the AGM battery is discharged instead to 80%, you will get around 800 charge cycles

- If you discharge a lithium battery to 0%, you will get around 5000 charge cycles
- If the lithium battery is discharged instead to 30%, you will get around 8000 charge cycles

Lithium batteries last on average, 4-10 times longer! And they have a larger usable capacity. This is why they are drastically cheaper than a sealed lead acid, in the long run. (Check out what batteries I recommend by visiting my website: http://www.mobile-solarpower.com)

Other Lithium Battery Benefits:

- Decreased Weight (usually 130-200% lighter)
- Decreased Size (usually around 50-70% smaller)
- Higher discharge and charging rates, and lower resistance. This means that all appliances will run with higher performance (But this depends on the battery! Cheap lithium batteries have limited discharge and charge current rates)
- Virtually maintenance free
- Better for the environment
- Does not give off dangerous fumes (Sealed lead acid batteries can technically discharge fumes, but only if they are used improperly. All lead acid batteries have the potential to gas fumes)

If you do not plan to power much and do not travel a whole lot, an AGM sealed battery will work and is the most common battery used for mobile solar power systems. If you plan to use the battery for many years and need something that will provide the best performance, a lithium battery is what you need.

Sizing up a battery

When you go shopping for a battery, you need to calculate its watt hour rating to see if it will provide enough storage for your needs. Unfortunately, batteries usually do not advertise their watt hour rating, but they do advertise their amp hour rating and voltage.

To determine the watt hour capacity of a battery from its amp hour capacity rating and voltage, multiply them together:

- A 12 volt, 250 amp hour battery can store 3000 watt hours
  (12 volts x 250 amp hours = 3000 watt hours)

- A 12 volt, 100 amp hour battery can store 1200 watt hours
  (12 volts x 100 amp hours = 1200 watt hours)

- A 6 volt, 225 amp hour battery can store 1350 watt hours
  (6 volts x 225 amp hours = 1350 watt hours)

- A 24 volt, 100 amp hour battery can store 2400 watt hours
  (24 volts x 100 amp hours = 2400 watt hours)

Battery selection example

Our example solar power system will require 2400 watt hours of usable power. We need to find a deep cycle battery that can fulfill this requirement.

Total required lead acid battery size to fulfill a 2400 watt hour requirement:

- A 12 volt, 400 amp hour lead acid battery bank will supply 2400 watt hours of usable power (but will have a total capacity of 6000 watt hours), which is perfect for our 2400 watt hour energy requirement

But guess what! 400 amp hours of the lead acid battery will weigh around 250 pounds! If you have a large vehicle, this is no problem. But if you have a small vehicle, you may need to downsize your battery bank.

Total required Lithium battery size to fulfill a 2400 watt hour requirement:

- A 200 amp hour lithium battery will supply 2400 watt hours of usable power

Unlike a lead acid battery bank, a lithium battery bank will be smaller, and provide more power.

A 200 amp hour lithium battery will weigh around 60-70 pounds, which is a lot less weight compared to the 270-pound lead acid battery.

2. Selecting Solar Panels

This is the easiest part of the entire system! Here are the key factors:

- Smaller solar panels are inherently stronger and thus, recommended if you mount them on the roof of a moving vehicle. 100 watt solar panels are the best size. They are strong, cheap, easy to find and easy to install. (Large solar panels are still safe to use, but I would not recommend using them. Use large panels for stationary solar panel arrays instead.)
- The 2 types of solar panels available for off-grid application are monocrystalline and polycrystalline. Monocrystalline is technically better due to higher efficiency, longer life span and ability to handle higher temperatures. Both types of panels work well though, so buy whatever is available. If you have severely limited roof space, or plan to use the solar panels for more than 20 years, then buy monocrystalline panels.
- The higher the efficiency of the solar panel, the better.
- Buy solar panels that come with MC-4 connectors installed.

The best metric I have used for choosing solar panels is customer reviews. If the panels are of lower quality, you will find out instantly in the review section. So to make your life easier, find the highest rated 100 watt solar panel, and you should be good to go!

Buy solar panels in multiples of 2. This will allow you to wire them in pairs, that are wired in series. If you buy 100 watt solar panels, buy 2/4/6/8 panels. This will make the process of wiring them together much easier.

Flexible Solar Panels

If you care about the aerodynamics and gas mileage of your vehicle or the total weight of your system, flexible solar panels will be better than glass solar panels:

- A typical 100 watt glass panel will weigh 16 pounds.
- A typical 100 watt flexible solar panel will weigh around 3-6 pounds!

They are lightweight and aerodynamic, but they have some downsides:

- They run hot. This means reduced life span and less efficiency when compared to glass panels. Some can fail in a matter of months! Be sure to buy one with a warranty.
- They can bend, but not a whole lot. Usually they cannot bend more than 30 degrees, unless they will be permanently damaged. It depends on the panel though.

If you buy a cheap flexible solar panel, it may overheat and cause a fire. There are quite a few online reviews that mention this.

Also, flexible solar panels cost more:

- A typical 100 watt glass panel will cost around 100-130 dollars
- A typical 100 watt flexible solar panel will cost around 170 dollars

If your roof is curved or you are unable to mount glass solar panels, you will probably prefer a flexible panel array. If you have a curved roof but want to mount glass solar panels, you might consider adding a roof rack.

Most flexible solar panel manufacturers will tell you that you can flush mount their panel. This is never true. You need to have some form of air flow under the panel. This is easy to do. I usually buy small thin dowels and slide them under the panel. This allows for air to flow around the solar panel. Flexible panels do not require much air flow, but they need some.

I recently had to return a flexible solar panel because it over-heated. It had high ratings online, and I thought it would last years. After only 6 months it failed! This is why flexible solar panel warranties are typically shorter than glass solar panel warranties. I have a couple flexible solar panels still, but they all have 25 year warranties, so I should be good to go.

Most people will do fine with a glass panel array. They work well, are super safe to use and really cheap. They can handle the elements and highway speed winds. They may be heavier, but a full set of panels will not weigh that much. Most flat roof RV's and most vans will do well with a 400 watt glass panel solar array. It will weigh around 60 pounds, which is no problem for most roofs and roof racks.

3. Selecting a Solar Charge Controller

In a previous section, we calculated the size of our charge controller. Now we will determine what kind of charge controller to buy. There are basically two kinds:

- MPPT (maximum power point tracking) Charge Controller: An advanced power modulation circuit that can find the highest efficiency battery charge rate with the solar panel power provided.
- PWM (pulse width modulation) Charge Controller: An off and on switch with limited capabilities.

The main difference here is the efficiency:

- The MPPT charge controller is around 95%-98% efficient.
- A PWM charge controller is around 60%-70% efficient

And cost:

- A 20 amp PWM controller is 15 dollars
- A 20 amp MPPT controller is 130 dollars

But the MPPT charge controller is still cheaper! Because it can produce more power with fewer solar panels. So the extra 100 dollars that you have to spend on the more expensive "MPPT charge controller" is instantly saved when you do not have to buy an additional 100 watt solar panel (typically 130 USD).

If you have 400 watt solar array on your roof, and you are using a PWM charge controller, it's only producing around 280 watts! That means that using a PWM charge controller requires you to spend hundreds more on solar panels that are not producing usable electricity.

Keep in mind that a vehicle has a finite space on the roof. Carrying solar panels on your roof that are not creating usable power is horribly inefficient.

There are fake MPPT charge controllers available online. These will be labeled as an MPPT charge controller, but are actually PWM controllers. Avoid these by reading the reviews, or see what I recommend at my website: http://www.mobile-solarpower.com.

Other Charge Controller features to look for

Temperature Compensation: Required for all systems. Determines the temperature of your battery bank with a small wired sensor, and allows your controller to charge your battery bank at a safe rate for the current temperature. This will prolong the life of your batteries drastically. If you have a lithium battery, you will need this and possibly a temperature cut-off if the batteries get too cold. If you are using a home-built lithium battery in freezing temperatures, you will need to buy a "Cold Charge Disconnect" system. If you buy an easy to use lithium battery from my website, it has an automatic temperature safety system built into it.

Large and Strong Input Terminals: If you look at a cheap charge controller, you will notice that the input terminals are small and weak. If you put the wire into the controller and you can wiggle the internal circuit board, you should avoid it. If the input terminal is a large screw that can fit your pinky finger, and the wire feels solid when you secure it, you found a good one.

Charge Controller Monitor: This allows you to see how many amps and volts your solar array is producing. I highly recommend everyone buy a controller that has this feature. It can also show how much power has been generated during the day and helps to troubleshoot problems.

Charge Profile Control (for advanced users only): Some charge controllers allow the user to modify the charge profile of the controller so that it can charge a variety of battery chemistries. If you wish to charge a home built lithium ion battery for example, or a not commonly used battery cell, you will need to buy a controller that allows you to manually edit the charge profile parameters. I do not recommend this for beginners!

4. Selecting an Inverter

Inverters can range in price from 20 dollars to 5000+ dollars. Your needs will determine the size and type of inverter. It will need a remote switch so that you can turn it on and off from the living area.

The largest determinant factor in choosing an inverter is how much you wish to spend. It is a good idea to buy the largest size you can afford. Most systems will require a 1000-2000 watt inverter.

- If you only need to power a laptop and a LED television, use a 500-750 watt inverter
- If you want to power a small blender and some power tools, use a 1200-1500 watt inverter
- If you want to power most household appliances, including a microwave or small air conditioner, a 2000+ watt inverter is ideal

I cannot think of a situation where you would need an inverter that is larger than 2000 watts unless you are running a large water pump, large microwave, multiple power tools, or multiple inductive loads.

Keep in mind that all induction loads, such as motors and microwaves, require an initial increase in power consumption at start-up. If you use a 1000 watt inverter to power a 1000 watt microwave, it will probably not work. This is because the initial power that the microwave uses is around 2000 watts. If you plan to run large induction loads, you need to double the required power (or triple if you can) of the inverter. You should also use inverter cables that allow for this short burst of required start up power.

You may be wondering why I recommend a 500-750 watt inverter for a laptop and television, even though they typically only require 200 watts total. This is because most inverters have an annoying little fan. It is best to buy an inverter that can run your appliances without using its cooling fan.

If you use too small of an inverter, it will be working really hard to supply a constant power. It will also give off a good amount of heat, which is an energy loss. It is not fun to use a small inverter and have it fail on you because it cannot handle a load (especially if you are powering something like a computer).

There are two kinds of inverters:

- MSW (modified sine wave) Inverters
- TSW (true sine wave) Inverters (sometimes called a "pure sine wave inverter")

An MSW inverter is typically much cheaper. A TSW is more expensive, but offers a few benefits:

- Safer to use with sensitive electrical devices such as computers and monitors.
- Inductive loads, such as motors, induction cooktops and microwaves, run more efficiently with more power.
- Reduced noise in the system. If you use an MSW, you will hear an annoying buzzing when using some of your appliances. If you try to run audio equipment, you will hear this buzzing even more. Using a TSW will fix this problem.

Most people can run a cheaper, MSW inverter and have no issues. They work well and are reliable. There are some appliances that will not work with an MSW, such as electric blankets, photocopiers, and specialized medical equipment. For most people, this is not a big concern.

5. Selecting Wire

The overall efficiency and performance of your solar system is dependent on the thickness and type of wire you choose. Many people underestimate the importance of wire selection and run into many problems.

The largest variable to consider when choosing a wire is the wire gauge size, which is the overall thickness of the wire. The thickness of a wire is dependent on:

- Length of the wire required for a specific application
- Amp load that the wire must carry

Typical wire gauge size for solar systems range from 0/4 gauge (the thickest) to 14 gauge (the thinnest). The most common wire gauge size to use in a system is 10 gauge wire.

If the wires in your system are not large enough to carry the power that your solar panels produce, you will have problems. This can cause a huge energy loss and can create heat in unwanted areas (which can cause a fire).

Electrical fuses are required for all electrical systems. They protect your system from short circuit damage, battery damage, and electrical fires. Fuses can only protect your system if you use the right gauge wire throughout your system. More on this later.

Guidelines when choosing wire:

- Thicker is always better. If you can afford it, always choose a slightly thicker wire than what is needed.
- The more strands, the better. Always avoid single strand, solid core wire in vehicles.
- Copper wire is ideal.
- Solar panel wire insulation must be UV resistant, and specifically designed for solar panel use. If you use regular wire that does not have UV resistant properties, the insulation will slowly degrade and crumble over time.

The longer a wire is, the thicker it must be to carry a load efficiently. Keeping solar system components close to each other will reduce the length of the wire required, which means higher efficiency and lower overall financial cost.

Wire can be expensive! Especially for large inverters where large wire thickness is required.

A trick to reduce wire thickness, without losing performance or efficiency, is to increase the voltage. The amount of current that a wire carries determines its thickness. If you have a 12-volt battery bank, you will not be able to increase the voltage (unless you put your batteries in series, which can give you 24 volts, which I do not recommend. It is technically better, but most appliances run on 12 volts, not 24 volts. It is much easier to just stick with 12 volts). But what you can do is increase the voltage of your solar panel array. This is easily done by wiring them in series. We will talk more about this in the installation section, but just keep in mind that increasing the voltage creates a smaller current, which means that you can safely use a smaller wire.

You can have a perfect wire, but if the connectors that attach the wire to the system are not properly crafted, you will encounter problems. We will go over this in the installation section. Just keep in mind that a wire is only as good as its connector. One bad connection can bottle-neck an entire system and create huge losses.

There are many different kinds of wire insulation. Many are designed for specific applications, such as high heat resistance, high voltage protection, UV resistance, oil resistance and more. For most solar system wires, any type of insulation will do fine, except for the wires that connect the solar panels to the charge controller. These need to be UV resistant and made for solar application.

The color of the wire should be consistent to avoid confusion. Typical color coding choice is using a black negative wire and a red positive wire. Some RV's will use black and white wires instead of the traditional black and red. Sometimes the black wire is the negative wire, and sometimes the white wire is the positive wire. This is why it is always important to use a multimeter tool to check the polarity of any questionable power wire.

General Wire Size Guidelines:

- Solar panel hookup wire, in most vehicle mounted solar systems, should be 10 gauge in thickness. If the wire length from the solar charge controller to the solar panels is longer than 25 feet, then 8 gauge wire is recommended. These recommendations apply for 17-24 volt output solar panels. If you put the panels in series which will increase the voltage to 34-48 volts, you can safely use a 12 gauge wire. 10 gauge wire is still more ideal. This wire must be UV resistant, so be sure to buy "solar panel specific" wire.

- Choosing a wire thickness to connect the solar charge controller to the battery bank can be tricky. This is because the voltage traveling through the wire is less than the solar panel wires, which means more current, which requires a thicker wire.

    If your solar charge controller is less than 10 feet away from your battery bank, use these recommendations:
    -20 amp charge controller requires 12 gauge wire
    -30 amp charge controller requires 10 gauge wire
    -40 amp charge controller requires 8 gauge wire
    -60 amp charge controller requires 6 gauge wire

    If your solar charge controller is further than 10 feet away from the battery, and less than 16 feet from the battery bank, use these recommendations:

    -20 amp charge controller requires 10 gauge wire
    -30 amp charge controller requires 10 gauge wire
    -40 amp charge controller requires 8 gauge wire
    -60 amp charge controller requires 4 gauge wire

    For my system, I have a 40 amp charge controller, 4 feet away from the battery bank, and I use 6 gauge wire. It is always best to use a slightly thicker gauge wire than required, to decrease losses.

- Inverter wires must be THICK. They need to carry more current than any other wire in a solar system. A 2000 watt inverter can use 200 amps! (sometimes more if you are powering an induction load) If you plan to run power tools, blenders, or a microwave with your inverter, you need a proper thickness wire for your inverter. The easiest way to choose an inverter wire is to buy it prefabricated. An inverter wire kit can be bought online for a very cheap price, and usually, comes with the connectors and a fuse already installed.

If you have a 1000 watt inverter, jump on the internet and search for a "1000 watt inverter wiring kit". If instead, you want to build your own wiring kit, this is what I recommend for inverter wires under 6 feet:

-250 watt inverter requires 12 gauge
-500 watt inverter requires 10 gauge
-1000 watt inverter requires 4 gauge
-1500 watt inverter requires 2 gauge
-2000 watt inverter requires 0 gauge
-2500 watt inverter requires 2/0 gauge
-3000 watt inverter requires 4/0 gauge

After you install your inverter, attach a large load appliance and feel the wires. If they warm up a little bit, that's ok. If the wires feel hot, you need a larger gauge wire.

- Wires that supply electricity to appliances are typically 12 gauge. This recommendation will work for any appliance that is 240 watts or less and is less than 13 feet from the battery. If the wire is especially long, but under 22 feet, you will need a 10 gauge wire.

If your wires are extremely long, or you plan to run a very large load, refer to the chart below:

12 Volt Wire Gauge Chart

Amperes	0-4 ft.	4-7 ft.	7-10 ft.	10-13 ft.	13-16 ft.	16-19 ft.	19-22
250-300	4-ga.	2-ga.	2-ga.	1/0-ga.	1/0-ga.	1/0-ga.	2/0-ga.
200-250	4-ga.	4-ga.	2-ga.	2-ga.	1/0-ga.	1/0-ga.	1/0-ga.
150-200	6 or 4-ga.	4-ga.	4-ga.	2-ga.	2-ga.	1/0-ga.	1/0-ga.
125-150	8-ga.	6 or 4-ga.	4-ga.	4-ga.	2-ga.	2-ga.	2-ga.
105-125	8-ga.	8-ga.	6 or 4-ga.	4-ga.	4-ga.	4-ga.	2-ga.
85-105	8-ga.	8-ga.	6 or 4-ga.	4-ga.	4-ga.	4-ga.	4-ga.
65-85	10-ga.	8-ga.	8-ga.	6 or 4-ga.	4-ga.	4-ga.	4-ga.
50-65	10-ga.	10-ga.	8-ga.	8-ga.	6 or 4-ga.	6 or 4-ga.	4-ga.
35-50	10-ga.	10-ga.	10-ga.	8-ga.	8-ga.	8-ga.	6 or 4-ga.
20-35	12-ga.	10-ga.	10-ga.	10-ga.	10-ga.	8-ga.	8-ga.
0-20	12-ga.	12-ga.	12-ga.	12-ga.	10-ga.	10-ga.	10-ga.

Length in feet

6. Battery Bank Voltage Monitors

Knowing the voltage of your battery bank will give you an idea of the current state of charge of the batteries and how much electricity you are drawing from the battery (voltage drops as amp load on the battery increases). Battery monitor guidelines:

- Wire it directly to the battery bank and nowhere else. Do not wire it to the solar panels or an appliance. Wire it directly to the battery bank with 14 gauge wire. This will ensure an accurate reading.
- Have the voltage monitor mounted in an easy to see location. Do not mount it in a closet or in a battery compartment. It should be clearly visible when you are using appliances that are connected to your system.
- Battery bank voltage is only an estimate of the batteries current capacity. Solar batteries usually have what's called a "surface charge" that can make it seem like your batteries are full. Because solar panels will charge your batteries consistently all day, the surface charge is present at all times except the morning, before the sun comes up. Just keep this in mind if you plan to draw a lot of electricity from your battery bank during the day time.
- The voltage reading will drop when the battery bank is supplying a load. In order to make an accurate "state of charge" reading, all appliances that are connected to the battery bank must be turned off.
- A fully charged lead acid battery bank requires a voltage reading of 12.7 to 14.5 volts. This is while the solar charging system is connected to the battery bank. If it is night time, and there is no solar power being supplied to the lead acid battery bank, a fully charged battery will read 12.6 to 13.0 volts. Lithium battery voltages are quite different and you will need to check the batteries manual.

How low can you safely discharge your battery?

This can be a difficult question to answer. If you have a sealed lead acid battery, the typical answer is 50%, which means 12.1 volts when all appliances are off. If you want your batteries to last for a very long time, you will want to discharge your batteries to 70%, which is around 12.3-12.4 volts. If you use the full depth of discharge of a deep cycle battery, which means discharging to 20% capacity or 11.7 volts, your batteries will degrade much faster.

If you have a lithium battery, the BMS (battery management system) will usually allow you to discharge the battery down to 0-20% and no more.

Also, if you do not use your battery or only discharge a little bit, such as 98% of the total capacity, other problems can occur. You need to discharge to at least 90%, once a week, to keep the batteries happy.

Long story short:

- If you have an AGM sealed battery, discharge to 12.2 volts and never go below 12.1 volts.
- If you have a lithium battery, check to see how the BMS works and try to discharge down to 20% capacity.

7. Fuses and Fuse Holders

A fuse is meant to be the "weakest link in the chain" of your solar system. If something in the system is drawing too much power, such as a short circuit (this occurs when the positive and negative lead makes contact), the fuse is designed to heat up and "blow". This will instantly disconnect the battery from the appliances and wires.

If a fuse is not installed in a system and a problem develops, the wire or an appliance will become very hot until it is destroyed. This can cause a fire.

A properly rated fuse, in a strategic location, will prevent most problems from happening.

But a fuse can only work if it is attached to a wire that is large enough to carry enough current to "blow" the fuse. If the wire is too small, or if the fuse is too large for a wire, the fuse will not blow. No longer is the fuse the "weakest link in the chain" of your system. The fuse will not blow, and the wires of your system will become the fuse. Because wires cannot blow like a fuse, they will instead become very hot, until something is destroyed.

If you have wired an appliance with an 18 gauge wire (which is quite thin and small), and it is attached to a 50 amp fuse, and a short circuit occurs in the circuit, there is a high likelihood that the fuse will not blow and that the wire will get excessively hot. This is because the wire itself is becoming the fuse.

Big Appliance = Big Wire = Big Fuse

Smaller Appliance = Smaller Wire = Smaller Fuse

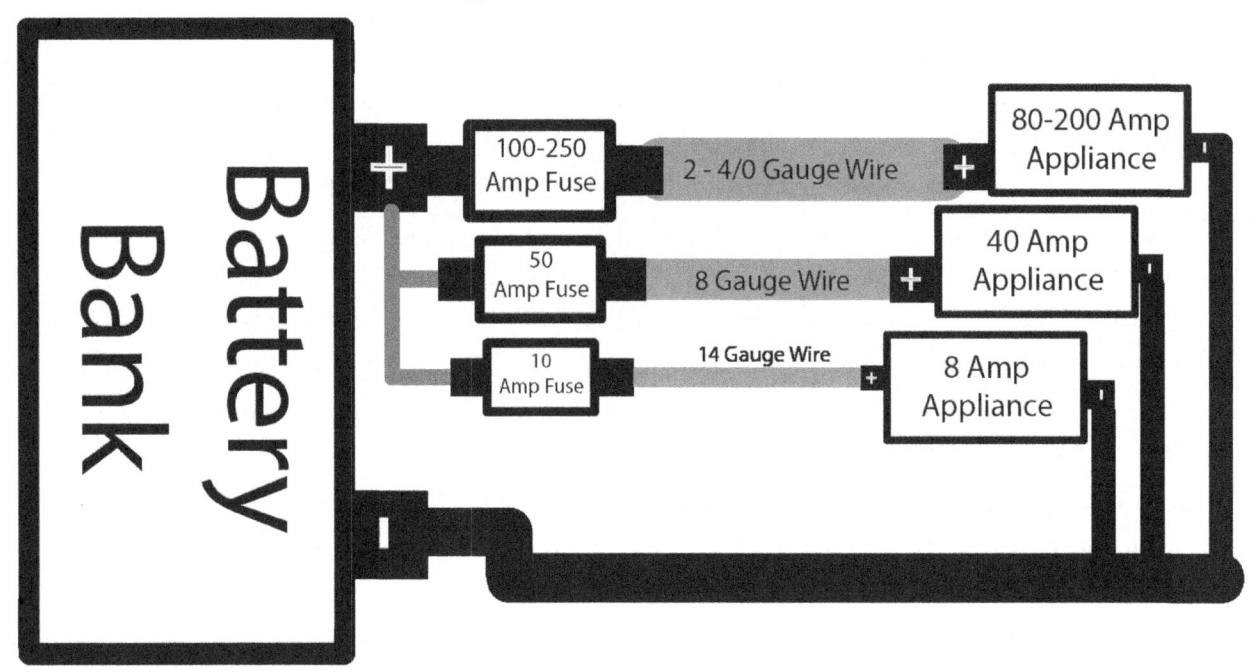

Choosing a proper gauge wire and fuse size is a pretty easy and straight forward task. Fuses are rated in amps, and if you know how many amps a wire can carry, you will know what type of fuse it needs. If you follow a few simple rules, you will be good to go.

Before we go over what fuse ratings to use, we need to cover what kind of fuse holders are currently available. Before you install a fuse into your system, you typically will install a fuse holder of some form.

There are three main types of fuse holders:

- **Bolt-on Fuse Holders and Fuses-** These attach directly to your battery bank terminals. Typical sizes are 50 amps to 250 amps.

- **Fuse Block-** A fuse panel where multiple small fuses supply power to appliances. Typical Sizes are 10 amp, 15 amp, 20 amp and 30 amp. Each fuse block can hold anywhere from 3 to 10+ fuses.

- **In-Line Fuse-** A fuse that is spliced into a wire. These fuses can be located anywhere, but are typically used for small appliances where the power wire is small. You can splice these in-line fuses into large wires, but it is not recommended. Common sizes for inline fuses for solar system application is 5 to 15 amps, but you can buy 30- 100+ amp sizes (but it is not recommended. A bolt-on fuse would be better to use instead).

How to Calculate the Fuse Size

Appliance amp draw x 1.25 = Fuse amp rating

- If an appliance draws 20 amps, use a 25 amp fuse (20 x 1.25 = 25)
- If you need to power an appliance that requires 5 amps at 12 volts, with the fuse block, you will need a 6.25 amp fuse. This is not a standard size, so a 7 amp fuse will work just fine. If the wires supplying the led strip are thick enough to carry 10 amps, then use a 10 amp fuse.
- If you need to power a fridge that requires 8 amps at 12 volts, to your fuse block, insert a 10 amp fuse into the fuse block where you will attach the positive lead of the fridge. Then connect the negative lead, or the ground, to the negative terminal of the battery.

Let's say that you have a small appliance that requires only 10 amps at 12 volts. Let's say that this small appliance has nice thick wires attached to it that can carry 25 amps (such as 12 gauge wires at 10 feet in length). In this situation, a 15-20 amp fuse would work extremely well.

The size of the fuse is determined more by the thickness of the wire than anything else. If you use a thicker gauge wire, you will be able to safely use a large fuse. (Unless the appliance is badly designed)

You may figure that a smaller fuse will be safer. In most instances, it is, but it can cause some problems. A fuse gives a good deal of resistance to a system, which means that some of your electricity will be transformed into heat. The fuse will literally warm up, all the time. This is a power loss. Using the properly sized fuse, and wire will be the safest and most efficient solution.

Important locations and ratings for fuses

1. There should be a large bolt-on fuse on the positive lead of your battery bank that connects to the inverter and fuse block, and any other large appliances you plan to connect to the battery bank. This is the main "safety fuse" that should only blow if there is a large short circuit close to the batteries. This fuse should be rated to match your inverter:

   -500 watt inverter should have a 50 amp fuse
   -1000 watt inverter should have a 100 amp fuse
   -2000 watt inverter should have a 210 amp fuse
   -3000 watt inverter should have a 315 amp fuse

2. After you attach the large bolt-on fuse to your battery bank, you can now add a fuse that will supply the solar charge controller. You can mount this fuse directly on the bolt-on fuse. This fuse should be rated slightly larger than your solar charge controllers amp rating:

   -20 amp charge controller wire will need a 25 amp fuse
   -30 amp charge controller wire will need a 35-40 amp fuse
   -40 amp charge controller wire will need a 50 amp fuse
   -60 amp charge controller wire will need a 75 amp fuse

3. Now we can attach the final fuse to the battery bank, which is the fuse block. This will house multiple fuses so that you can power your appliances. If you would like to wire the fuse block directly to the large bolt-on fuse that is located on the battery bank, you need to use a thick wire. This wire should be at least 6 gauge. This should be enough to trip the large bolt-on battery fuse if it were to short circuit.

   If your fuse block is small, and cannot handle a 6 gauge wire, you will need to fuse the wire that supplies the fuse block. A 50 to 100 amp fuse will be ideal, but check the instructions of the fuse block to see what it recommends.

8. Other Power Sources

Solar power is the ideal method of electrical power creation, but not the only one. Depending on your location, you may want to supplement your system with other power sources.

Shore Charging (plug in chargers)

Shore charging refers to charging your battery bank while parked, with an AC power outlet. If you are stationary and have an AC outlet nearby, you can design a pretty powerful system with less solar panels by using a plug-in charger. But you will be dependent on having an AC power outlet nearby.

There are quite a few battery chargers to choose from today, but the largest determinant factor is how many amps the charger can produce. Common amp outputs are 15/30/45/100 amps.

Many of the battery chargers come with alligator clamps. This works well for emergencies, but if you plan to use it a lot, you need to cut the alligator clamps off and crimp a connector so that you can permanently install it to your battery bank.

A large amp output charger should attach to your battery bank on the large bolt-on fuse. If you are using a very tiny battery charger, you can wire it to your fuse block. Example: If your battery charger produces only 15 amps, wire it to a 20 amp fuse on your fuse block.

Be sure to buy a 12-volt charger! There are also 24-volt and 6-volt chargers available. Make sure that your charger is designed to charge the batteries you are using.

Generators

Many people, including myself, are getting tired of generators. They smell, make noise, consume petroleum products and require mild maintenance. But they work! If you plan to run large loads, such as power tools, an electric oven, or large water pumps, you may need to supplement your system with a generator.

Most generators cannot be wired directly to the battery bank. You will need some form of charger, such as a shore power charger mentioned a second ago. You plug your battery bank charger into the generator and you are done!

Most large RV's will have a storage compartment specifically for a generator. If it doesn't, you will probably need to add a generator mount for the bumper. If you have a scooter/motorcycle rack, you can mount the generator there. Be sure to bolt it down, and lock it to your vehicle's chassis with a chain so that no one steals it. Do not mount the generator on the roof, or near the living area. Be sure to operate it in an open area, to avoid breathing the fumes.

Wind Turbines

These work well, but typically not on vehicles. They require being folded away while you drive. If your vehicle is stationary and located in a windy location, then set up a turbine! They can produce a lot of power and are pretty easy to setup. Typically you just bolt it to something secure and wire it to your system. Some turbines come with a voltage regulator, and you can hook it directly to your battery bank.

How to Install a Solar Power System

First, install the:

1. Batteries
2. Solar Charge Controller
3. Solar Panels

1. How to install a battery bank

Look around your vehicle and think of a place to mount your battery bank. RV's and some vans will have a battery bank compartment. Some vehicles will require you to build your own compartment. Battery banks typically favor:

- A dry location that is protected from the rain.
- A location that will protect the battery terminals from being tampered with.
- A location that prevents the batteries from tipping over while you drive.
- An insulated location that will prevent large temperature fluctuations. You may need to insulate your battery bank compartment.
- Idealistically the compartment should be made of an insulator to prevent accidental short circuits. Plastic battery boxes are great for this purpose.
- If you have sealed batteries, the location does not require ventilation. But ventilation is always nice to have. A sealed battery will not give off fumes if it is used properly, so usually you do not need to worry. If your batteries are vented, your battery compartment will need ventilation.

Battery banks are heavy. You must store the batteries somewhere between the front and the back tires. If you load the passenger side of your vehicle down with a battery bank, you must do the same for the driver side. Use the location of the tires to draw an imaginary "X" in your vehicle, and try to situate the batteries in this location.

If you have really tall batteries, you may need to secure them with a strap, so they do not tip over. Most deep cycle batteries have a large footprint and are so heavy that they do fine on their own without any special mounting methods.

Once you find a location, put them there. As long as it is a relatively safe location, you shouldn't have any problems.

2. How to install a solar charge controller

Most solar charge controllers are designed to be mounted on a wall. This is because the controller has cooling fins, and it requires some form of convective ventilation to cool it down. When you mount it on a wall, make sure that the space above and below is clear, so air can pass through.

The hardest part about choosing a charge controller location is finding a wall location that is close to the batteries. The closer the controller is to the batteries, the better.

You can screw the charge controller into the wall with some screws, or you can use mounting tape (if the charge controller is relatively light weight). Either way works, so use whatever is easiest. If you have a large charge controller like an MPPT charge controller that is larger than 40 amps, you will need to reinforce the wall that you mount it on or mount it to a piece of wood, then mount the piece of wood onto the wall.

3. How to install the solar panels

Solar Panels require:

- Air flow under the panel
- Secure, waterproof connection to the roof

Here are your options:

- **Option 1:** Check out the "Solar Panel Mounts" section on my website at http://www.mobile-solarpower.com for a constantly updated list of methods, pictures, hardware and adhesives for attaching your solar panels to various roof materials.
- **Option 2:** Use a standard solar panel mounting kit to bolt the solar panels directly to a roof rack. If you can install a roof rack to your vehicle, this is your best option!
- **Option 3:** Use a set of "drill-free corner mount solar panel brackets" and some VHB tape or weather-proof construction adhesive to mount your solar panels. Your roof has to be flat where you put these mounts. I list these on my website and they are my favorite!
- **Option 4 (not recommended):** Secure the panels directly to the roof with a solar panel mounting kit. Use long bolts and fender washers to secure the solar panel mount L-bracket's to the roof. Next, use caulk to seal all of the holes.
  Having holes in the roof of any vehicle is a bad idea. Also, the bolts do not look that pretty on either side of the roof. Most people will buy a manufacturer recommended mounting kit. Usually, these kits come with special L-brackets that will secure your solar panels to your roof with large screws. Usually, these L-brackets are plenty strong, but the roof that they attach to can be weak. Many RV roofs are made with relatively weak materials, such as fiberglass and foam, and are not designed to have solar panels mounted on them. Most solar panel mounting L-brackets are designed to be attached to bare wood or bolted to metal. If your vehicle has a metal roof, you can drill into that, but you may not want to do that.
- **Option 5:** If you have a curved roof that is made of fiberglass and you do not want to drill into it and you cannot add a roof rack to your vehicle, you probably need flexible solar panels. You can still use 3M-VHB mounting tape, or whatever the solar panel manufacturer recommends, but this may be your only option.

Solar Panel Safety Line

Attaching heavy glass solar panels to the roof of a moving vehicle is generally a bad idea. To reduce the possibility of having a solar panel fly off and kill someone, you should add a safety line.

How to add a safety line:

1. Drill a hole into the frame of every solar panel in your array. Be sure to not damage the solar panel while drilling the hole. Most 100 watt solar panels come with small holes that work well for this purpose.
2. Find a weatherproof, UV-resistant rope or cable to attach the solar panels together. A thin stainless steel cable is ideal, but a marine grade rope can work as well.
3. Attach the solar panels to each other with the safety line. You can have them connected individually or together like a spider web. It doesn't really matter. Just make sure that if a solar panel flies off, it is attached to another solar panel.
4. Attach the safety line to a sturdy object on the roof of your vehicle. This could be a roof rack, RV roof ladder, Air Conditioner or even an unused TV Antennae.

Now that you have a solar array safety line, you need to take extra care to not trip on it while you are walking on your roof. This can be done by securing the safety line slack to the roof with roof sealing tape. If you have a lot of solar panels on your roof, try to mount them along the perimeter of the roof, so you are less likely to trip on a safety line or solar array wire.

This safety line can save lives, so it needs to be perfect. Spend the extra money for better supplies and ensure that your safety line will work in the event of a solar panel mount failure.

Should you tilt your solar panels?

Usually, no. It is better to mount them so that they face parallel to the roof.

If you live far from the equator, tilting your solar panels may be required. In this situation, you may consider mounting them on the side of your vehicle. You can also use some flexible solar panels to make a large blanket of panels that you can mount on the side of your vehicle when needed.

Tilting your panels does increase overall output when the sun is low, but it's just difficult to do on a moving vehicle. If you use a solar panel tilt mount, you will have to fold it up every time you drive.

You can always wire up a linear actuator to an ignition circuit so that the panels tilt when you are parked, and fold up when you drive. But then you will have to wire up a motor control and light sensing circuit to direct where your panels should tilt. If any of these parts fail, you may have a solar panel fly off your roof and kill someone.

So to avoid problems, mount your panels parallel to the roof of your vehicle.

How to Wire up your Solar Power System

1. Learn how to crimp and select wire
2. Connect all batteries and add a main fuse
3. Connect solar charge controller to the battery bank
4. Connect individual solar panels together to create a solar panel array
5. Connect the solar panel array to the solar charge controller
6. Connect the inverter and fuse block to the battery bank
7. Install a battery monitor
8. Connect appliances

Crimping is the method of connecting the components of your solar system to the wires with small metal fittings called "crimp connectors". If this is your first solar system, you will need to teach yourself how to crimp.

To master this skill quickly, I recommend jumping on the internet and watching a video tutorial. This is a hands-on activity that requires watching someone else do it, and a couple minutes of practice. Or you can follow the steps below:

Crimping Summarized

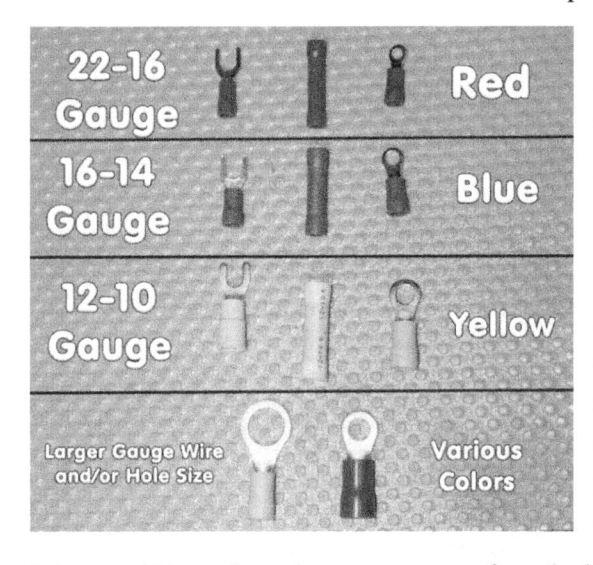

Crimp connectors come in standard sizes, which are red/blue/yellow, and a variety of other sizes for specific applications.

Crimp connector sizes are determined by:

- Gauge of wire it is attaching to
- Size of bolt or terminal (size of the hole)

Let's say you need to connect a 4 gauge wire to a 5/16" inch diameter battery terminal (read battery manual to find this measurement). Go to your local hardware store and find a crimp connector made to work with 4 gauge wire that has a 5/16" hole. That's pretty much it. As long as the wire gauge and the hole size is correct, the crimp terminal will work.

It is a good idea to buy a large assortment of standard size crimp connectors before you begin.

If you need a specialty crimp connector, such as battery connectors that require a specific diameter hole, you can find them at a local hardware or automotive parts store. If you need a large crimp connector for connecting an inverter to a battery, you may need to special order it (or buy prefabricated inverter cables).

Tools Required for Crimping:

- **Crimp Tool**: The crimp tool below is designed to crimp, and it can also cut large wires. The small indentations under the cutter blades are what you will use to crimp. One indentation is made to crimp connectors with insulation, and one indentation is made to crimp connectors without insulation. This type of crimper typically has really long handles so that you can get some serious leverage when you crimp.

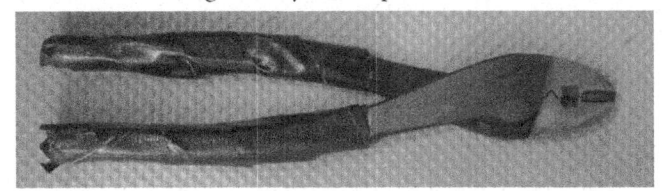

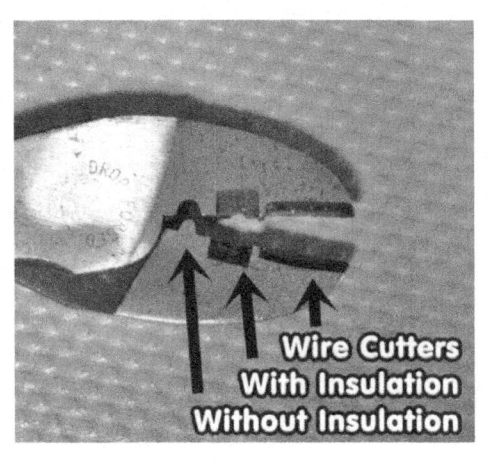

Another option is to buy a ratcheting or hydraulic crimp tool. As long as it is specifically designed to crimp, you are good to go. You should never use pliers to crimp a connection!

Note: There is one type of wire crimper that is just awful! It is designed to crimp wires, but it is the cheapest kind around, and will not work well. Avoid it:

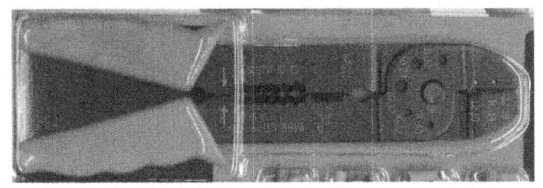

- **Wire Stripper**: There are a few different kinds available, but the one below is my favorite. The stripper below also comes with a crimp tool (small indentations near the handle), but do not use them! The leverage of a wire stripper is not large enough to crimp properly. Only use a dedicated crimp tool for crimping. Use the wire stripper only to strip the wires.

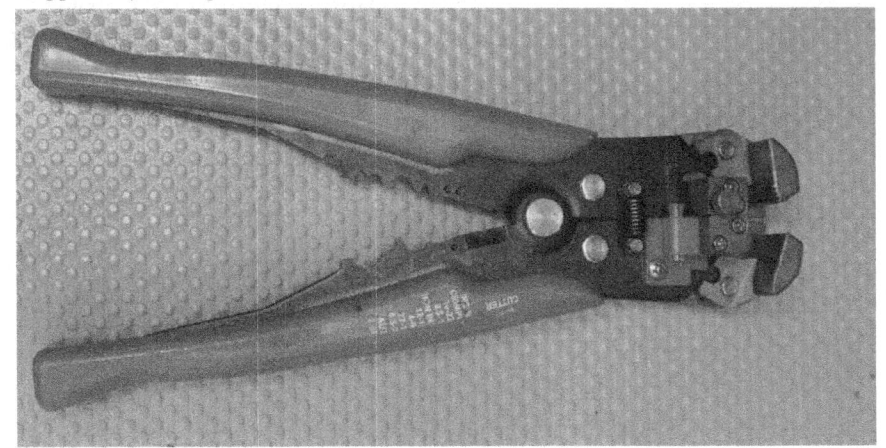

- **Clean, dry and oil-free workspace:** Wash your hands or wear gloves. Wires and connectors need to be clean before they are crimped.
- **Rubbing alcohol and towel:** to clean dirty wires.
- **Heat Shrink Insulation:** not required, but recommended. Electrical tape works well too.

How to Crimp

Step 1: Select a proper size connector for the gauge of wire that you are using

Step 2: Strip the wire so that the wire can fit inside of the connector. Be sure not to damage the wire strands while you are stripping the wire. Ensure that the wire is clean before and after you strip it.

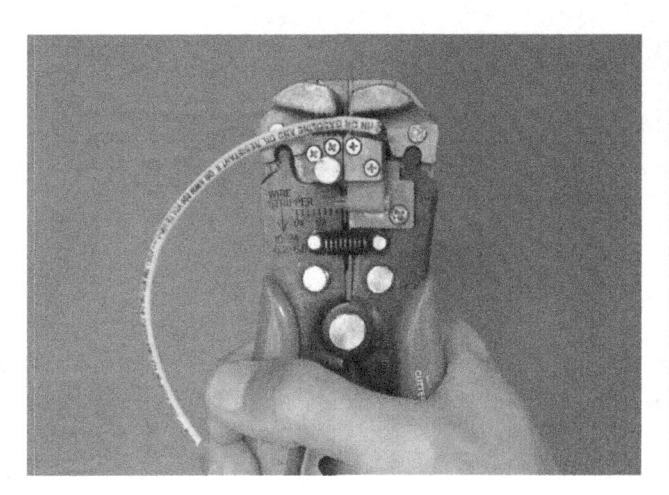

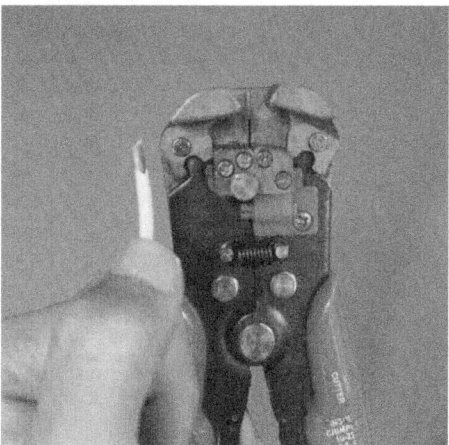

Step 3: Insert the wire into the connector to ensure a snug fit. If you cannot easily slide all the strands of wire into the connector, you have the wrong size. If the wire sticks out too far on either side of the connector, you will need to re-strip the wire so it fits perfectly into the connector.

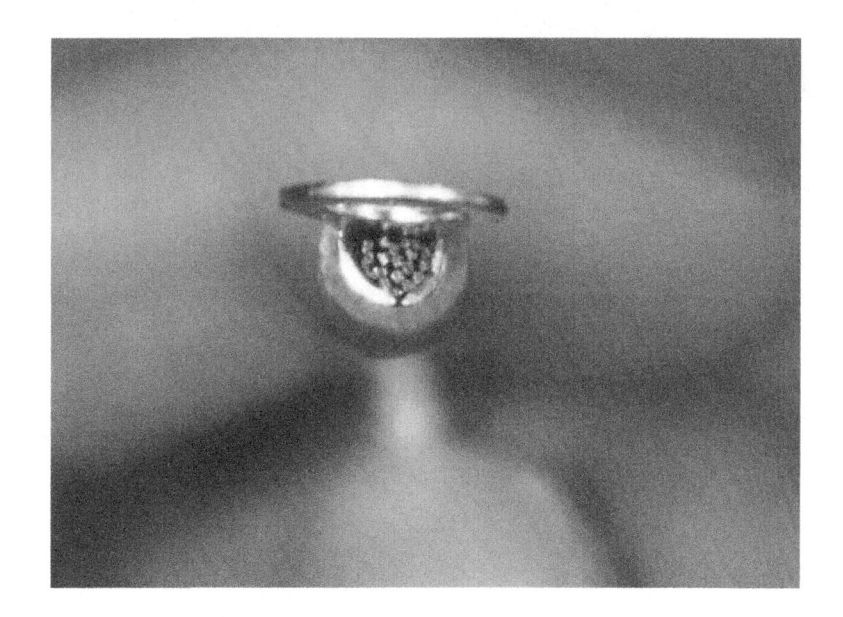

Step 4: Remove the wire from the connector, and insert the connector into a crimp tool. Slightly squeeze the crimp tool to hold the connector in place (do not squeeze hard!) Notice that the connector pictured below (small circle in the crimp tool jaws) has insulation, so it is mounted in the appropriate indentation. If your crimp tool has small dots that are blue, yellow and red, then match up the connectors color to the proper color coded indentation.

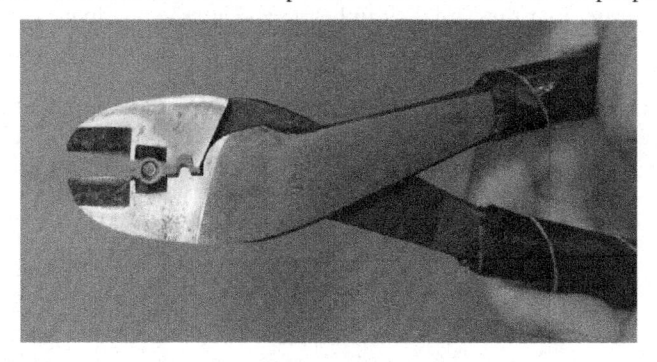

Step 5: Insert the wire and then squeeze the crimp tool with a tight grip. When the connector is smashed, and the wire is secure, it is considered a "termination". The wire and the connector become one piece of metal when a full termination occurs.

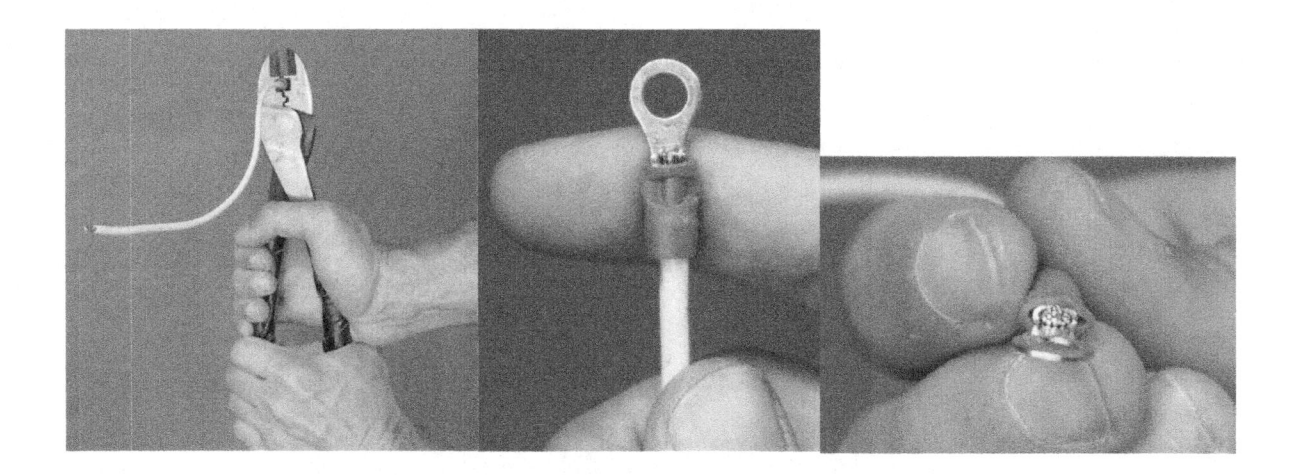

Step 6: Inspect the connection. If you pull on it, it should feel solid. If it is a weak connection and feels loose, or the wire falls out, then you will need to re-strip the wire and start over with a new connector. Some connectors also have a second area to crimp, which will hug onto the wire insulation.

Step 7: Add heat shrink if necessary. Heat shrink will help to prevent stress on the joint. If you are crimping large wires, such as for an inverter, heat shrink is highly recommended. If you have never used heat shrink before, do not be scared! It is easy to use. Cut a small piece and cover the connector's insulation and part of the wire. Then use a heat gun or stove to slowly heat up the heat shrink till it shrinks around the connector. Heat shrink is recommended for crimp connectors that do not have insulation, but I like to add heat shrink to all connectors for added protection.

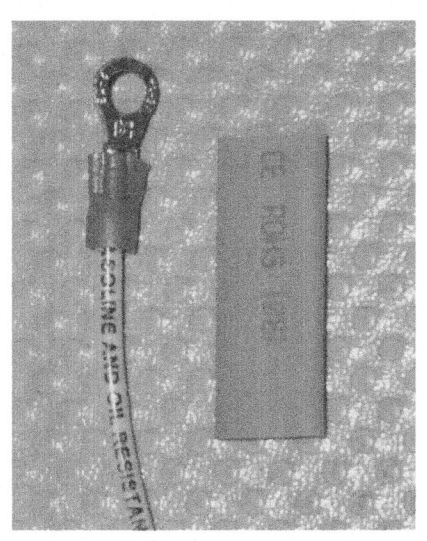

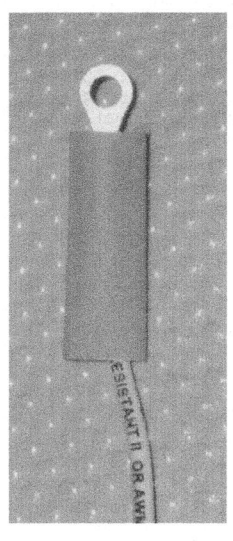

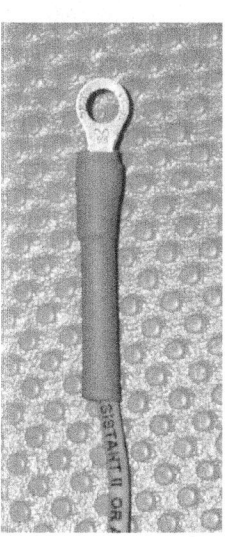

Crimping seems so simple, but many people fail to create a strong connection between the connector and the wire. This can happen for a few reasons:

- **An improper tool used for crimping:** many people use pliers to crimp a connector. This will not create a complete termination and will result in a horrible connection. You must use a crimp tool and nothing else. Do not ignore this tip! You need to buy the proper tool, or crimping is a waste of time!
  If you plan to crimp a large connector, you will need a special crimping tool. Most people do not have this on hand, so call a local car stereo shop or electronics store to see if they have one that you can borrow instead.
- **Improper size connector:** all connectors are rated for a specific gauge of wire and no other. If your connector is too big or too small, you will have a bad connection.
- **A dirty wire:** some wires that you try to crimp may be covered in dirt, water or oil. If the wire is not completely clean, a complete termination will not occur and you will have a weak joint that can fall apart. You will also experience an energy loss. Use rubbing alcohol and a small towel to wipe wires down before you crimp them.
- **Allowing the wire's insulation to go inside of the connector:** This will block the wire from merging with the connector, which will prevent a termination.
- **Using multiple wires in a connector:** or folding the wire in half to make it fit into a larger sized crimp connector.

If you watch a video tutorial online and read the steps above, get out some wire and give it a go! Do a few practice crimps, and you should be ready to wire your solar system.

1. Connect all batteries and add the main fuse

If you have multiple batteries, you will want to wire them together before adding the main fuse. If you have 12-volt batteries, wire them in parallel. You can attach 12-volt batteries together with 2-4 gauge (or larger) battery hookup wires which are available at most auto parts stores.

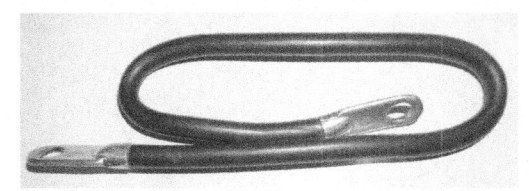

Parallel Battery Bank Configuration

These are short wires that come with a prefabricated connector that connect each battery to the one next to it. Connect the positive terminals to the other positive terminals, and the negative terminals to all the other negative terminals.

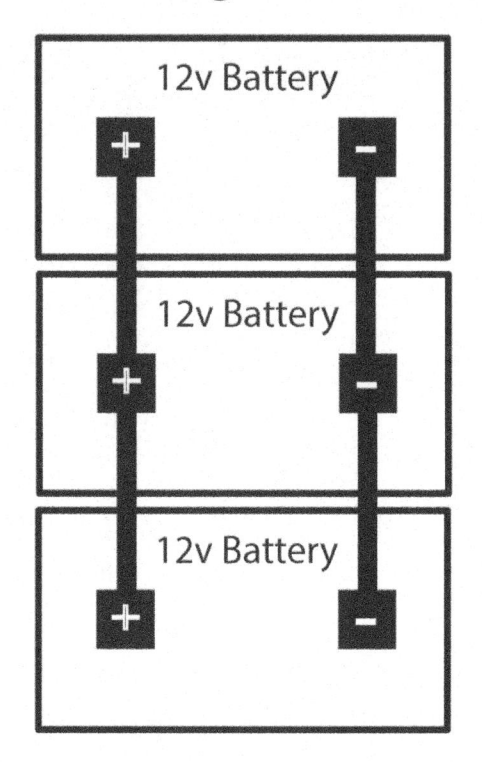

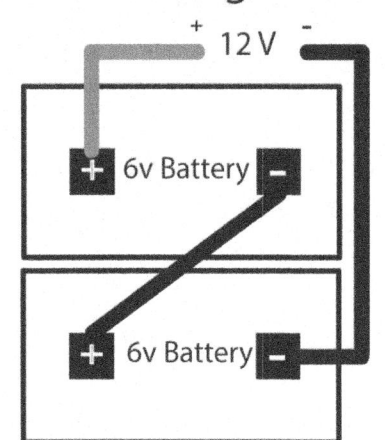

6 volt Battery Bank Series Configuration

If you are using 2x 6-volt batteries, wire them in series. This will give you a total volt rating of 12 volts.

Series and Parallel Configuration for 6x 6 volt batteries

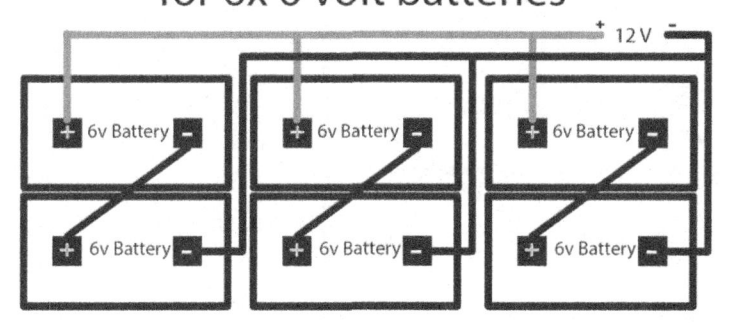

If you plan to add more 6-volt batteries in the future, or you plan to use 4x 6-volt batteries, treat each pair of 6-volt batteries the same as a single 12-volt battery, and wire them all in parallel.

Bolt-on Fuse Location for Parallel Wiring Configuration 12 volt Battery Banks

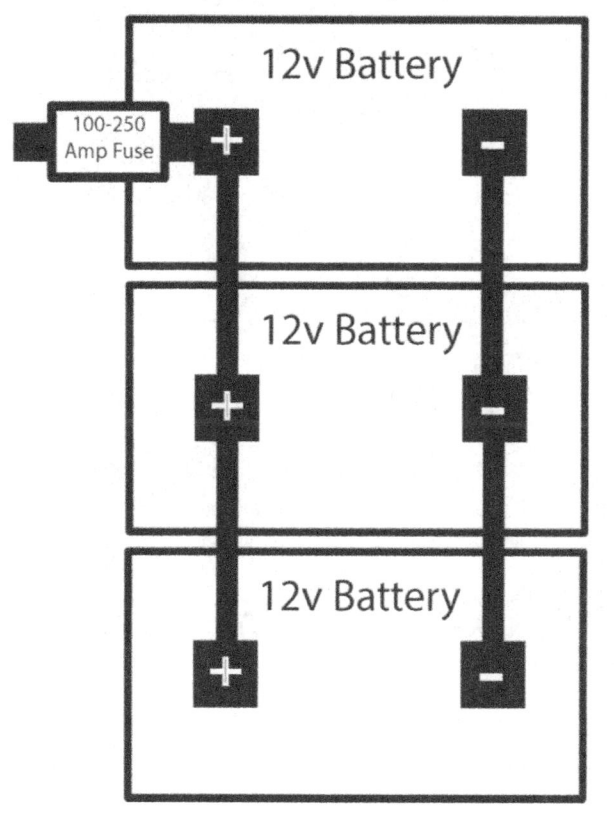

Choose a positive (red) battery terminal in your battery bank that you plan to be closest to your inverter, charge controller and fuse block. This is where we will install a bolt-on fuse. Wipe the chosen battery terminal with a towel and make sure it is clean, then install the bolt-on fuse.

Check to ensure that the bolt-on fuse is designed for use with your battery terminal. They should fit snug, and have plenty of surface area where they connect to each other.

Bolt-on Fuse Location for multiple 6 volt Battery Banks

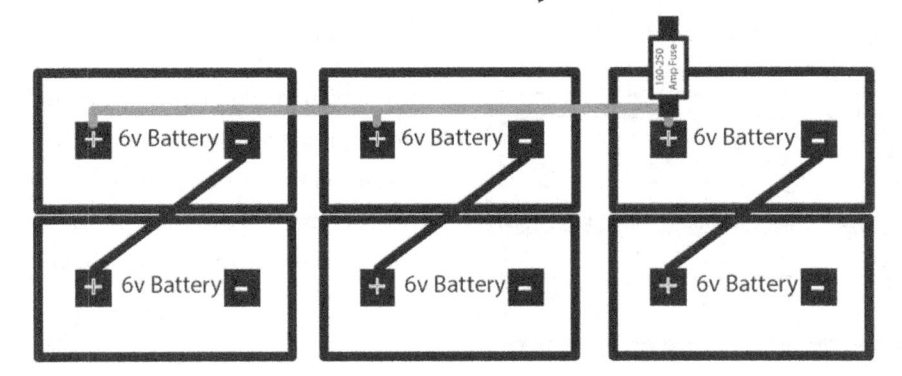

If you plan to use multiple 6-volt batteries, attach the bolt-on fuse to one positive terminal of one of the battery banks.

What it should look like

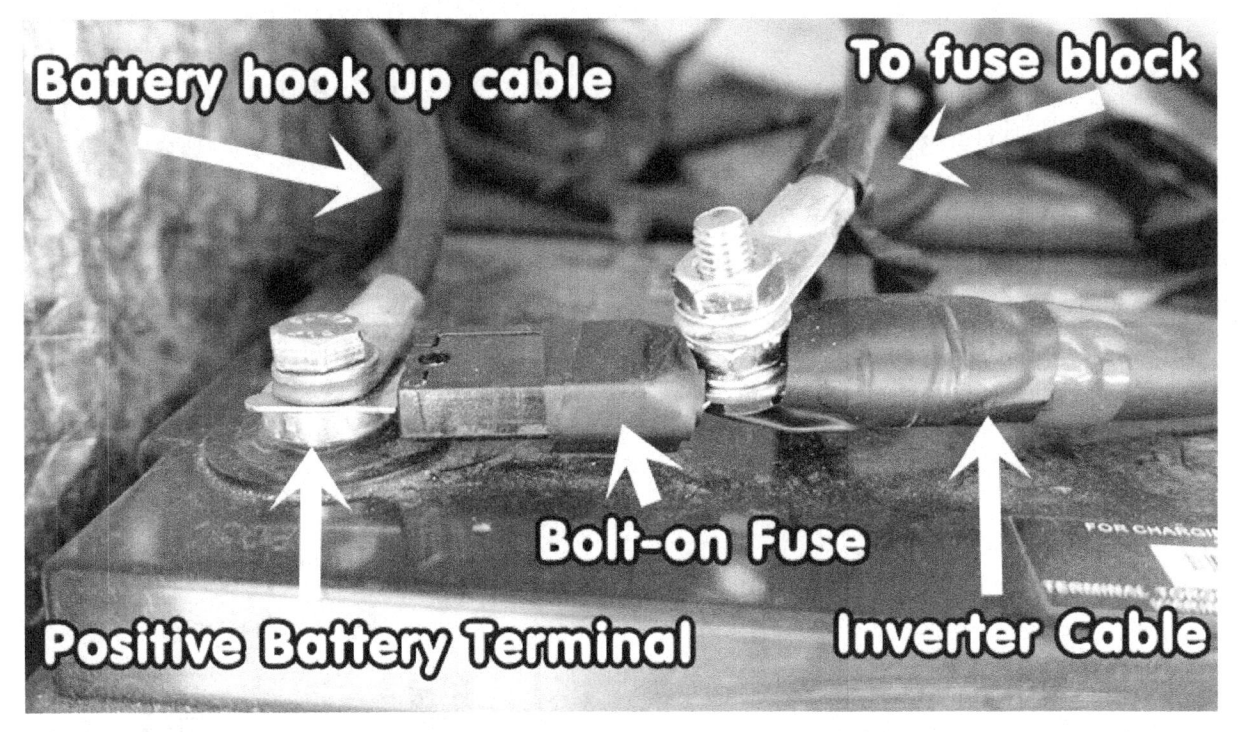

The "battery hook up cable" in this picture attaches to another 12-volt battery's positive terminal.

You do not need to worry about hooking up the inverter or fuse block yet, but this is what it should look like when you do.

Tighten the battery terminal bolt with a wrench, but be sure to avoid creating a short circuit across the two battery terminals. If a metal tool, which is a conductor, touches both battery terminals, the tool will heat up and you will also create a large spark. This can damage the battery, and hurt you.

After you have installed the bolt-on fuse, find a way to cover and protect the positive battery terminals. This will prevent accidental short circuits in the future. There are rubber terminal covers available, but if you cannot find one, use electrical tape. If you have multiple batteries, cover all of the positive terminals.

When you are done wiring the batteries and adding a fuse, you have created your battery bank! Now we can connect the battery bank safely to the rest of the system.

2. Connect solar charge controller to the battery bank

If you do not want to damage your controller, you must wire it to the battery bank first and the solar panels second.

Typically 8 gauge wire is used if the charge controller is 40 amps or less, and located less than 10 feet away from the battery. If it is a large charge controller or further than 10 feet, refer to the wire gauge chart in the previous wire gauge section. I personally like to use 6 gauge wire here for a 40 amp controller. This is overkill, but this is your batteries largest energy source. You want minimal wire loss here. The larger the wire, the better.

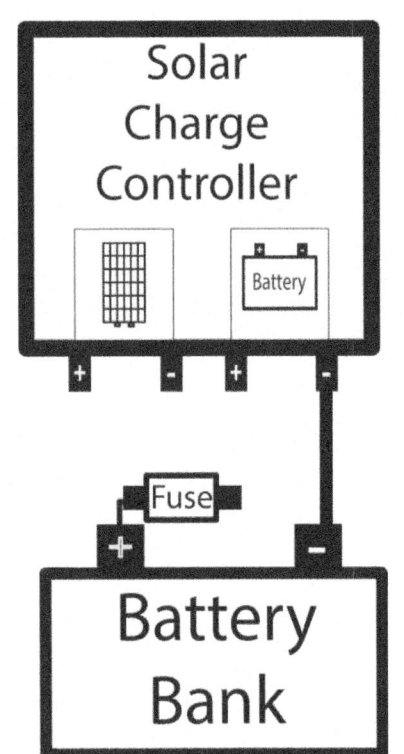

The first wire we need to install will start at the negative battery bank terminal, and will connect to the solar charge controller terminal that is labeled "negative battery". (See picture on left)

Strip one end of a wire and attach a crimp connector that can attach to your batteries negative terminal. Bolt this wire and connector to the negative terminal of your battery bank. Next, run this wire to the solar charge controller, strip it, and connect it.

Most solar charge controllers come with a screw down terminal. What this means is that the charge controller will have a small hole where you can insert a wire and it will have a screw which will secure the wire.

Use a screwdriver or Allen wrench to secure the wire in the terminal. Be careful not to over tighten this screw, especially if it is small.

If the wire is too small for the terminal hole, you can strip off more insulation, fold the wire in half to make it thicker, and reinsert it into the charge controller terminal.

Next, strip and connect a wire to the solar charge controller's positive battery terminal.

Run the positive wire alongside the negative wire that we just installed a second ago, and back to the battery bank area.

Now that the positive wire is near the battery bank, we can attach it to the battery bank fuse. Use a crimp connector that can attach your bolt-on fuse to your wire.

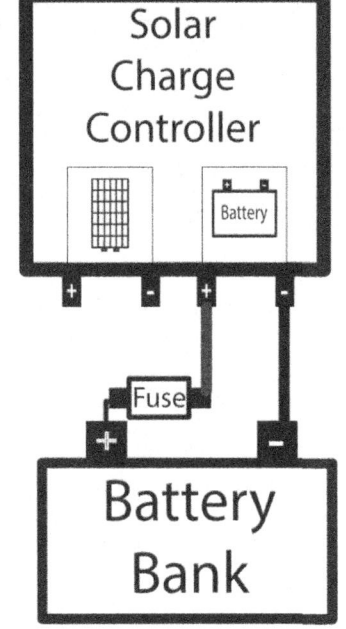

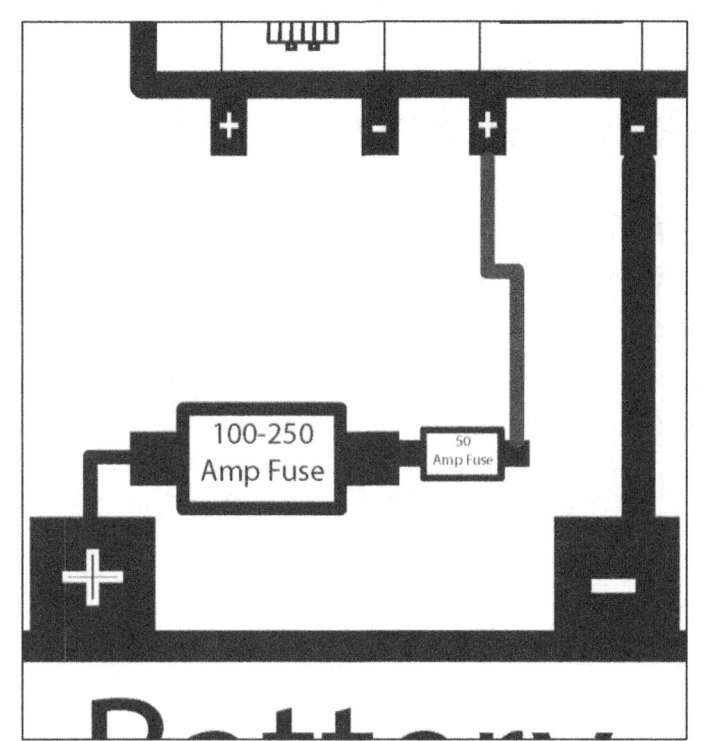

Depending on the size of the bolt-on battery bank fuse, and the gauge of wire you use, will determine where you need to attach the positive wire. If you have a large battery bank fuse, such as a 250 amp fuse, you cannot connect an 8 gauge solar charge controller wire to it. The wire is not thick enough to carry enough current to trip the fuse. The 8 gauge wire could potentially cause a fire.

So what you need to do is add an inline fuse, or a smaller bolt-on fuse to attach to the large battery bank fuse.

If you have a 40 amp charge controller, use a 50 amp fuse. (Pictured on left) Bolt a 50 amp fuse onto the big fuse, or attach it to another positive battery terminal on another battery. Either way works.

After you attach the positive wire, go back to the charge controller to see if it is "turned on". You should see a green light or a screen turn on! If it turns on, you are good to go. Good job! If it doesn't turn on, you probably switched the positive and negative wires. This can cause permanent damage to most solar charge controllers.

3. Connect individual solar panels together to create a solar panel array

You have two options for wiring solar panels together: in Series or in Parallel.

Solar Panels in Series

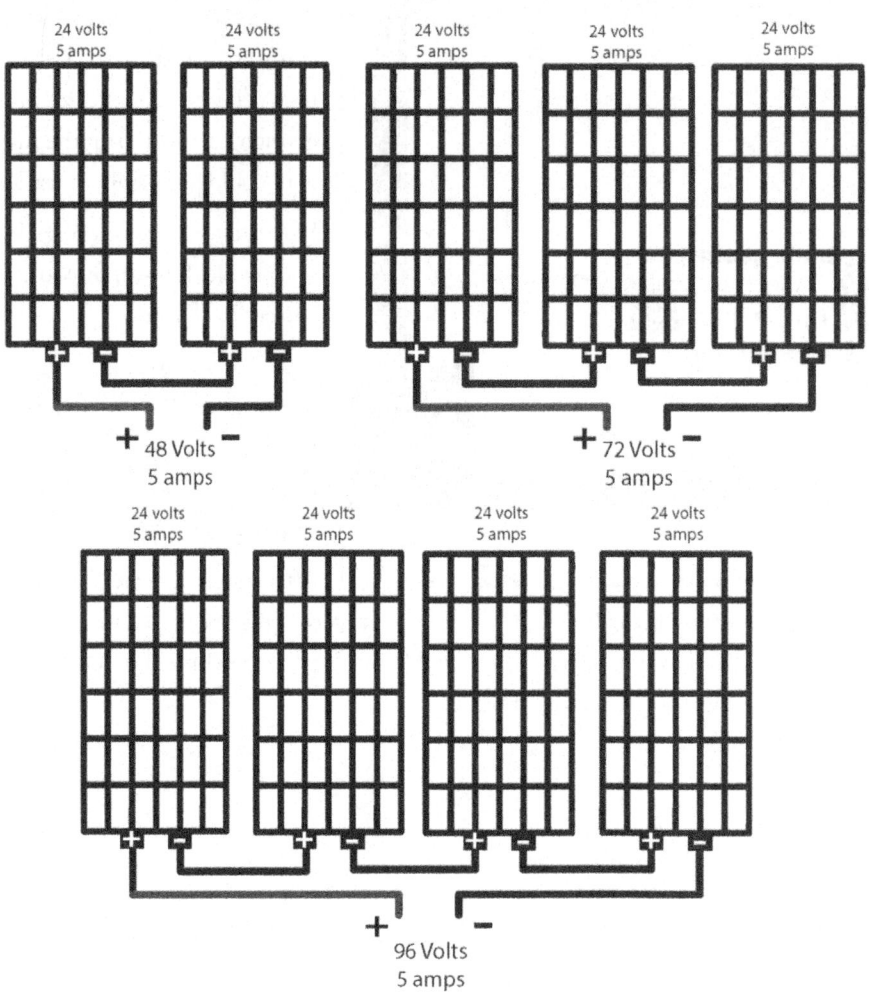

Pros of Series Configuration:

- The voltage will increase, and the amps will not change. This will produce fewer amps traveling through the wire, which will translate to less wire loss and increased efficiency.

Cons of Series Configuration:

- When a group of solar panels is wired in series, they act as one panel. Sunlight must shine equally on all panels to produce electricity. If one solar panel is covered from sunlight, power creation from all panels is severely limited.
- Solar panels must be of the same make, age and amperage rating, to wire them in series. They also need to be close to each other and angled at the sun in a similar fashion to work properly.

Solar Panels in Parallel

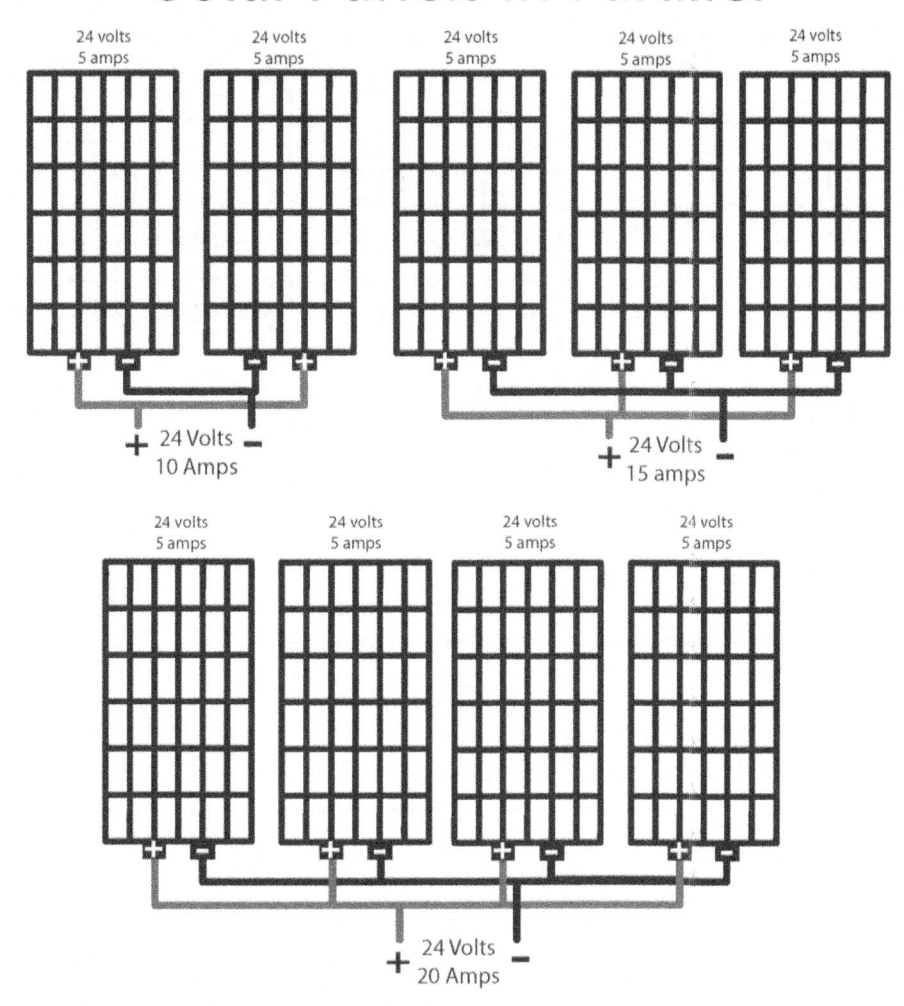

Pros of Parallel Configuration:

- Each solar panel produces power independently in this configuration. If you destroy or shade 1 solar panel, the other panels will still produce electricity.
- Panels can have different amp ratings, and it will be ok to connect them in parallel. If you have different sized panels that produce the same voltage, and you wish to combine them all, parallel configuration will work best.

Cons of Parallel Configuration

- When solar panels are wired in parallel, the voltage will not change, but the amps will increase. The increased amps will be harder to carry in the wire. You will require a thicker wire unless it will be horribly inefficient. If the wires are not thick enough, they will heat up and you will have a power loss.

The best configuration is a combination of series and parallel. Combine solar panels in pairs to produce a larger voltage to increase solar array efficiency. Connect pairs of panels, or "in series groups", together in parallel so that your system has some degree of redundancy if panels falter or are shaded from the sun.

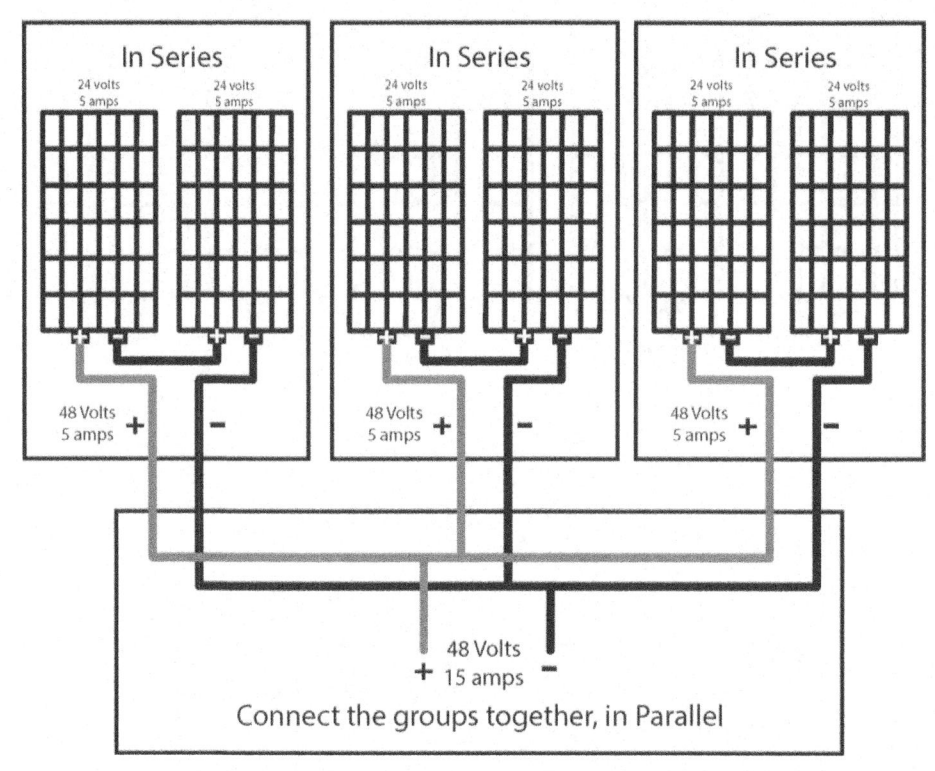

Connect the groups together, in Parallel

If you wish to add more panels to this array, buy 2 more solar panels and make a new "in series" group.

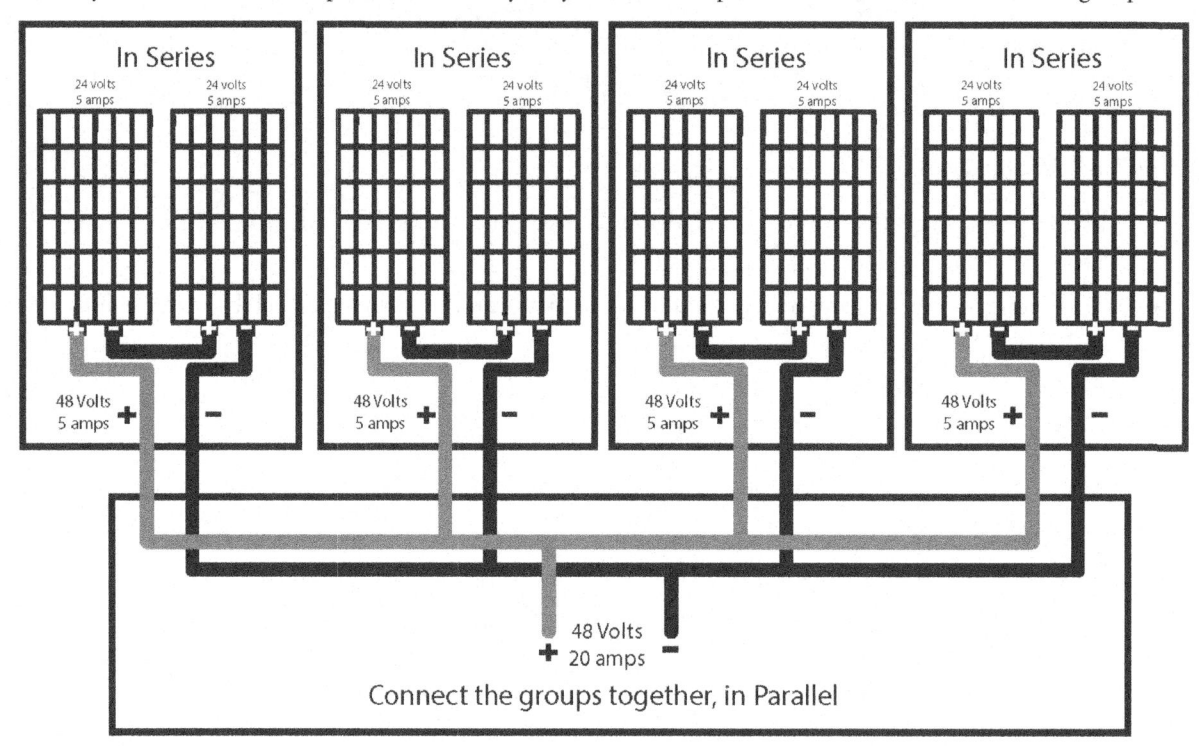

Connect the groups together, in Parallel

Wiring 2 panels for each "in series group" is ideal. If 3 or 4 panels are wired together in each "in series group", problems may occur. This is because all of the panels in a series configuration depend on each other to produce electricity. If one of the panels is shaded, the whole "in series group" will suffer (this is why it is so important to mount "in series groups" of panels close to each other and at the same angle towards the sun).

When you put solar panels on a vehicle, you never know where the sun will be. You may have to park under some trees where half of your solar panels are shaded, and half are exposed to the sun. In this situation, having all of your solar panels in series will severely limit the energy production of your system. So the benefits of a series configuration are great when you wire them in small groups, such as 2 panels, sometimes 3.

If you are building a stationary solar system and you have no shading problems, then you should wire the panels in series to create the largest voltage possible (that is safe to use with the solar charge controller).

If you wire up more solar panels in series, you may have a larger voltage, but the benefits of a higher voltage will not matter as much if your wires are thick enough to carry the amps supplied at lower voltages. Most solar panel hook up wires are 10 gauge in thickness. Pushing 48 volts through this wire will not be much more efficient than pushing 100 volts through this wire (even if your charge controller can handle 100 volts).

This is why a combination of series and parallel works best for mobile systems. We need to series connect in pairs to increase the voltage a little bit which will increase efficiency, and parallel connect these pairs together so that the array will still produce electricity if part of your vehicle's roof is shaded.

If you are using MC4 connectors, wiring your panels together in different parallel/series configurations will be a quick and straightforward process.

Connecting 2 solar panels "in series" is done by connecting one panel's negative wire to the other panel's positive wire.

If you wish to parallel connect multiple "in series groups", you will need to buy an MC4 branch connector. This will allow you to connect wires of the same polarity together as one.

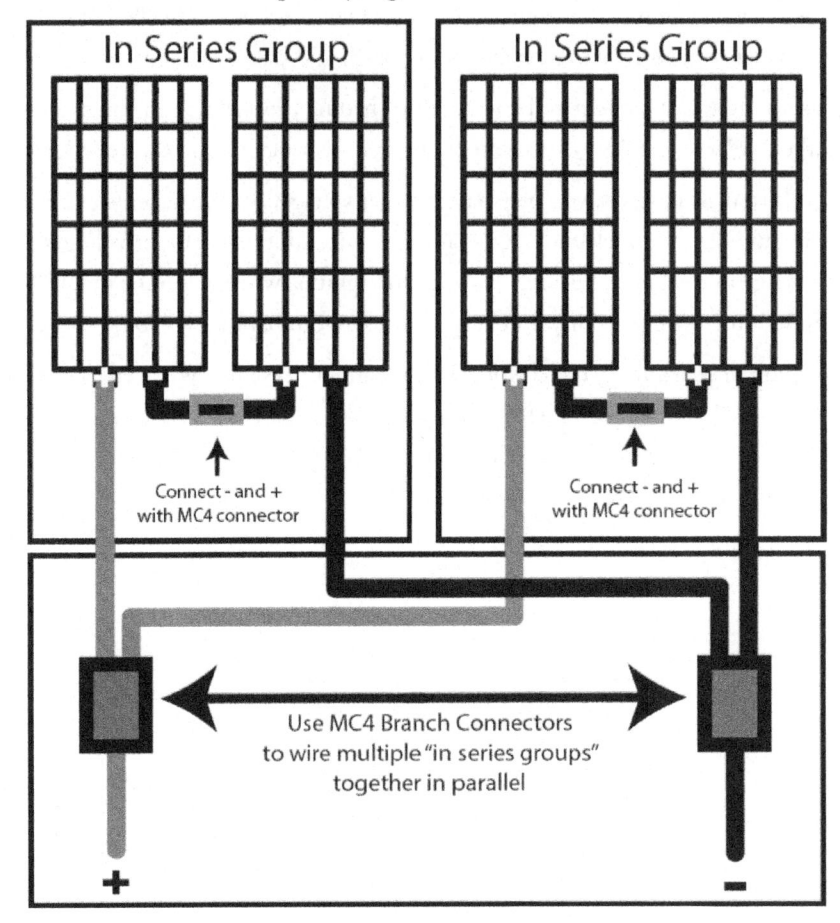

MC4 branch connectors come in different sizes. The size will determine how many groups you can connect.

Connect all of your panels together so that you have only 2 wires: one positive, and one negative that will connect to your solar charge controller. If you are wiring a lot of solar panels together, you will need to tidy up the cables with some zip ties.

Once your solar panels are mounted and connected together, your solar panel array is now complete!

4. Pass solar array wires through the roof and connect them to the solar charge controller

Typically this requires drilling two small holes into the roof to allow entry of the solar cables. To seal the holes, you can use caulk. Or you can use a "Cable Entry Gland" and some self-leveling lap sealant which is mounted on the roof of your vehicle where you wish to pass the wires through. They look great and will protect the 2 small holes from rain.

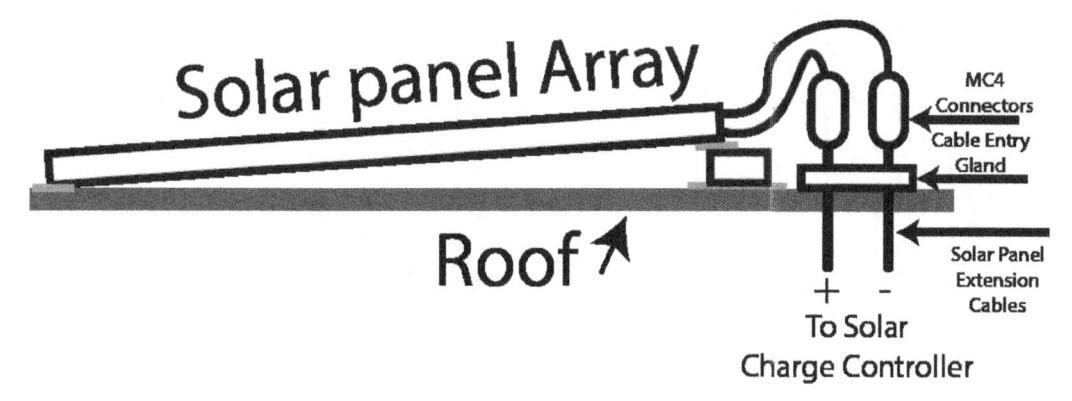

It is wise to buy solar panel extension cables. They connect the solar panel array to the solar charge controller and are designed to be passed through your roof with the "Cable Entry Gland". It is preferable to have all MC4 connectors and branch connectors located on the roof. MC-4 connectors should never be inside your vehicle.

Before you attach the solar panel array to your solar charge controller, you must ensure that the voltage produced by the solar panel array is not larger than the voltage that the solar charge controller can handle.

To double check the solar panel array voltage, use a voltmeter such as the one found on a multimeter to check the output voltage of your solar array. Be sure to expose the solar panels to sunshine when you check the voltage.

Also, check the polarity of the panels (which wire is negative or positive). Most multimeters will tell you if the polarity is correct with a minus sign in front of the voltage. This will tell you that the polarity is reversed. Refer to your multimeter's manual to be sure.

If you check the voltage of your solar panel array and it is not what you anticipated, you will need to rewire the panels.

If your solar array is producing the voltage that you calculated, and the polarity of the wires is correct, you can now connect the solar panel array to the solar charge controller.

Connect the negative solar panel array wire to the negative solar input terminal on your charge controller. Most solar charge controller terminals require you to strip the solar panel wire, insert it into the terminal then use a screw to secure the wire in place.

Next, connect the positive solar panel wire to the charge controller. This is what it should look like:

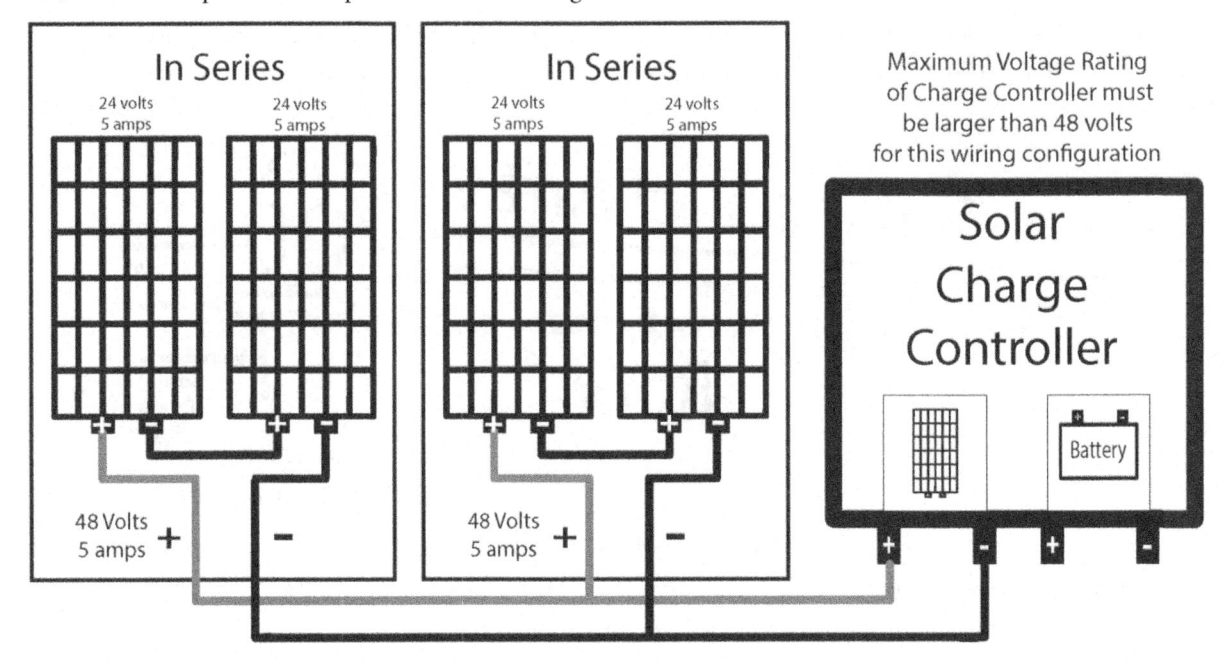

Now check the solar charge controller to see if it is receiving power from the solar panels. Typically it will have a green indicator LED and a small picture of a solar panel next to it. Each charge controller is different, so be sure to read your charge controller manual.

Your battery bank should now be charging with electricity produced by the solar panels.

Some solar charge controllers will require you to tell it what kind of battery you are using. Be sure to read the controller's manual to see if this is required. Now is the time to program this setting.

5. Connect the inverter and fuse block to the battery bank

The wires that attach to these components are typically the largest wires in a system and will supply power to all appliances. The fuse block will supply 12 volts DC, and the inverter will supply 110 volts AC.

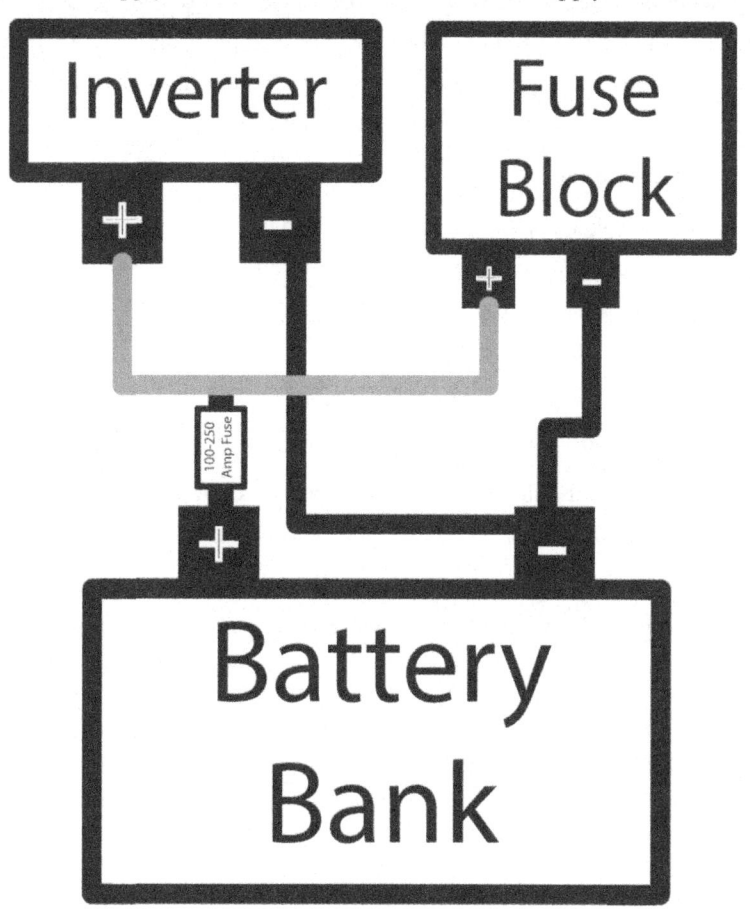

Fuse Block Installation

Usually, a fuse block will have a positive and negative terminal that you will attach directly to your battery bank. Typically a 4-8 gauge wire is recommended. Find a crimp connector that will fit nicely with the terminals on the fuse block and your battery bank bolt-on fuse. You can mount the fuse block anywhere, but mounting it in an easy-to-see location, next to the batteries is ideal. If you want the fuse block in the living area of the RV, so that appliances are easy to wire, you will need to use a pretty thick wire to supply the fuse block. Fuse block wire gauge thickness varies depending on how much power the fuse block must supply. Read the fuse block instructions to see what it recommends.

Inverter Installation

Inverters require THICK wires. A typical 2000 watt inverter runs well on 0 gauge wires. If you want to save yourself a lot of time, simply buy an inverter wiring kit online. This makes the whole process much easier. But if you want to buy the wire yourself, and add crimp connectors, you can do that as well.

When you crimp a large inverter cable, you will need a special type of crimp tool. These are found at car stereo shops and hardware stores in the electrical section. They are pretty simple to use. The hardest part of crimping a large wire is cutting the insulation so that you do not destroy any of the delicate wire strands in the wire. Crimp the connection the same as you would crimp a smaller crimp connection.

Online videos are very important for this! Watch a 3-minute video tutorial if you want to see it done.

If you cannot find a large crimp tool, and you have large wires and crimp connectors, you can put the wire into the crimp connector, slam it with a hammer, and then heat it up with a torch and feed solder into the connection. This is not as good as a crimp connection, but it will do the job if you are desperate. This connection can fail if it gets too hot. The solder would melt and the wire could slip out.

If you cannot find inverter cables at local stores, buy some battery cables instead. Most cables are only 4-6 gauge in size, but for smaller inverters, they work well.

The bolt-on fuse on the battery bank should allow for an inverter to be wired directly to it. You bolt the positive inverter cable to the fuse, and then the negative inverter cable, and you are done. It may spark when you attach the inverter and this is normal.

What can be difficult is running the AC wires out of the battery compartment, and into the living area. Or to install a remote switch to the inverter. You will have to get creative here and try to work with the design of your vehicle.

You may want to connect the inverter AC output to your vehicle's pre-existing AC system. This can be a little dangerous, so I recommend seeing a professional electrician to have this done. If you are new to this, run some AC extension cables for now.

6. Battery Monitor Installation

Now that your battery bank has a solar charging system, we can now add a battery monitor. The best battery monitor will be the "shunt". This is a special type of current meter that tells you exactly how much electricity is going in or out of your battery at any moment. Most shunts typically have a voltage monitor, so this is a great way to also estimate the state of charge. If you have a lithium battery, you do not need to worry about this metric (just make sure that the lithium battery BMS has low voltage protection). If you have AGM sealed lead acid batteries, you will need to keep an eye on the voltage while running large loads or when using moderate sized loads for prolonged durations.

Some shunts can be wired to tell you how many watts of power your solar panels are producing. Depending on where you mount your shunt will determine what the shunt will measure. If you attach it to the wire that supplies the solar charge controller, then you will be able to see how much electricity your solar panels are producing. You can instead wire the shunt to the fuse block, or the inverter, to see how much electricity those appliances are consuming.

A shunt attaches to the negative terminal of a component at the battery. Most shunts can be bolted to a negative battery terminal, so it is easy to add it on any system. It will take only five minutes to add it.

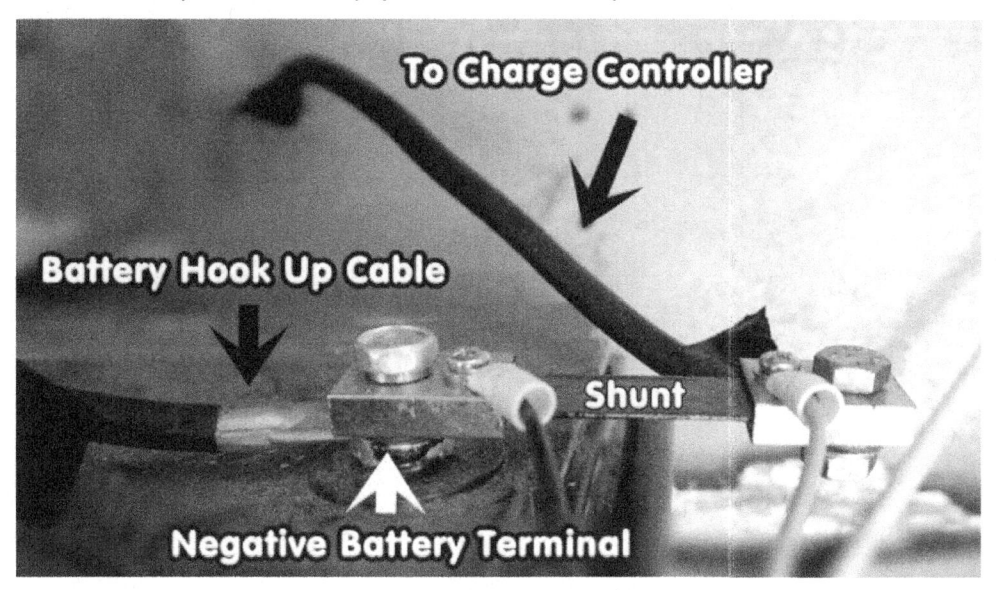

Keep in mind that if you ever disconnect a battery bank cable, you need to disconnect one of the solar panel wires from the charge controller. The controller does not like to be disconnected from the battery while still connected to the solar panels. This can cause damage to the controller. So before you install a shunt, be sure to disconnect one main solar panel wire from a terminal on the solar charge controller, and then install the shunt.

To summarize the typical shunt installation procedure:
1. Disconnect a solar panel main wire from a terminal on the solar charge controller.
2. Loosen a negative battery terminal and install the shunt directly to the battery. Then install the components that you want to measure, to the shunt. If you want to measure how much power is coming from your solar panels, wire only the solar charge controller to the shunt. If you want to measure the inverter, wire it to that instead.
3. Install small wires to the shunt. Usually, you will have 2 wires attached to the shunt itself, and 2 other small wires, a negative and a positive. These wires will run to a small LCD screen that will display the data that the shunt is observing.

Each shunt will be different, but that's the typical steps. Be sure to read the manual before you install it!

Adding DC 12 Volt Appliances

When you wire a DC 12-volt appliance to your solar system, you have three options:

1. Hard wire the appliance directly to the fuse block. This works well for stationary appliances.

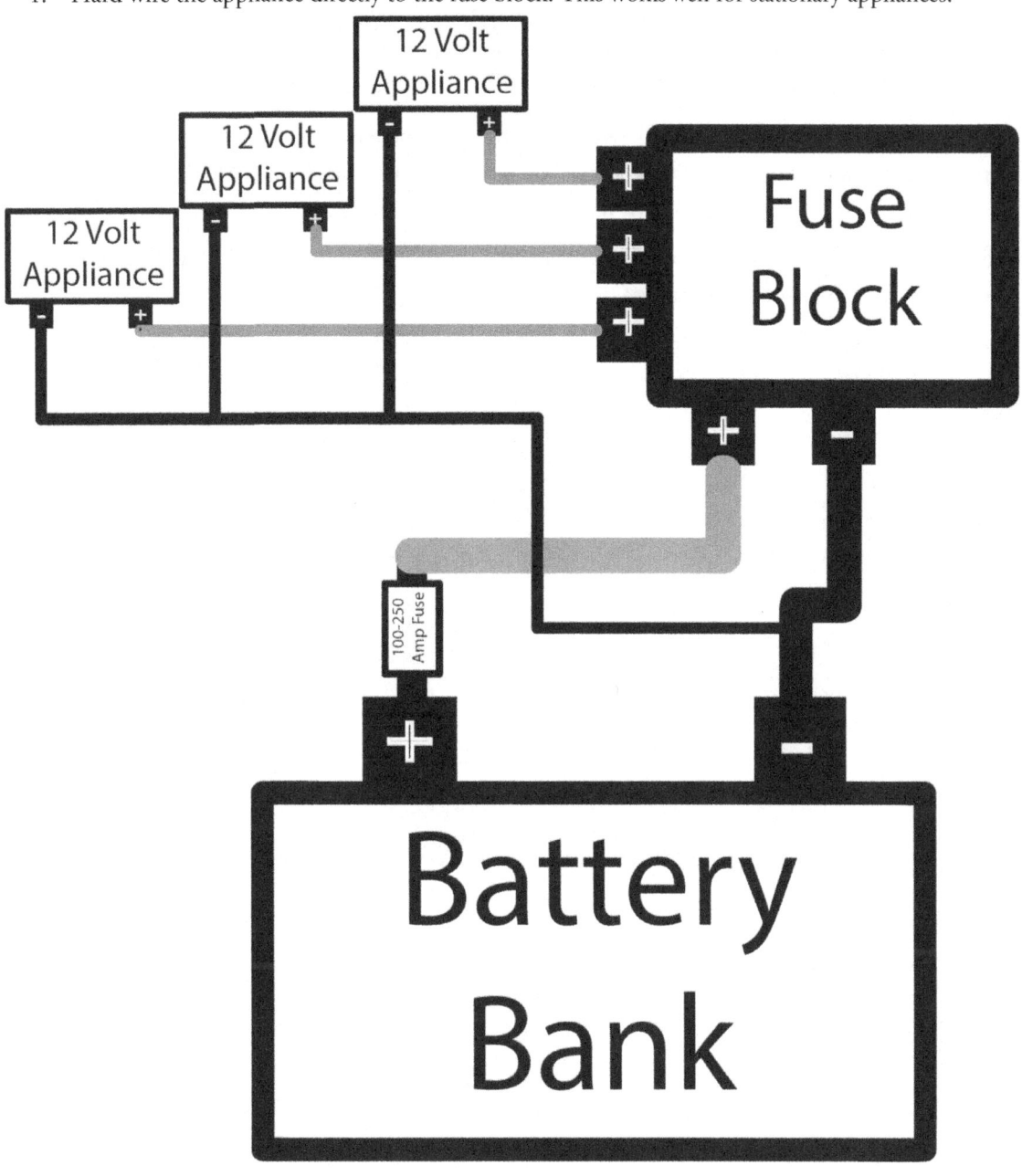

2. Wire the appliance to a male type plug connector, such as an XT-60 plug or Anderson Powerpole connector. This will allow you to connect the appliance only when you need to use it. This requires hard wiring a female type plug connector to the fuse block. You can wire multiple female plugs, and mount them in different locations in your vehicle.
3. Buy a cordless appliance that comes with a rechargeable battery, and recharge the battery with a 12-volt charger.

When you connect an appliance to a 12-volt system, be sure that it is rated for 12 volts! If you do not check, and it is rated for a lower voltage, it will be destroyed.

Depending on how you use an appliance will obviously determine how you will wire it. Typically, if you use it every day, and require it to be as efficient as possible, you will want to hard-wire it directly to the fuse block. If you do not use it every day, or if it needs to be portable, then a plug connector or cordless appliance will make sense.

Make sure that you use the proper gauge wire. I cannot say this enough times. Check the wire and fuse section if you want to know exactly what size to use.

If you want to wire a small appliance to your fuse block, you can repurpose an AC cord. Cut the ends and splice it in.

XT-60 connector

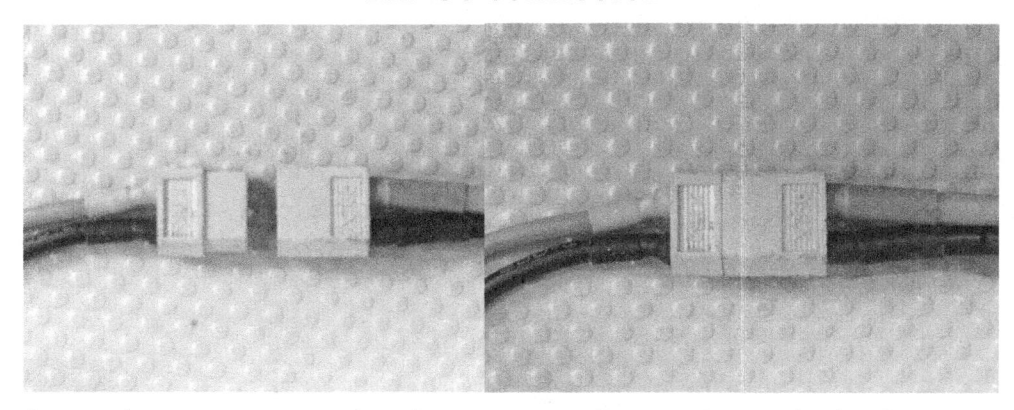

This is my favorite plug connector to attach appliances to a 12-volt system. It is rated to handle 60 amps and is super durable.

1. Wire and mount a few female type XT-60 connectors into strategic locations in your vehicle
2. Attach your appliance's power wires to a male type XT-60 connector

This way you can plug your appliance in at different locations, and put the appliance away when you are not using it. Wire the female connectors directly to the fuse block:

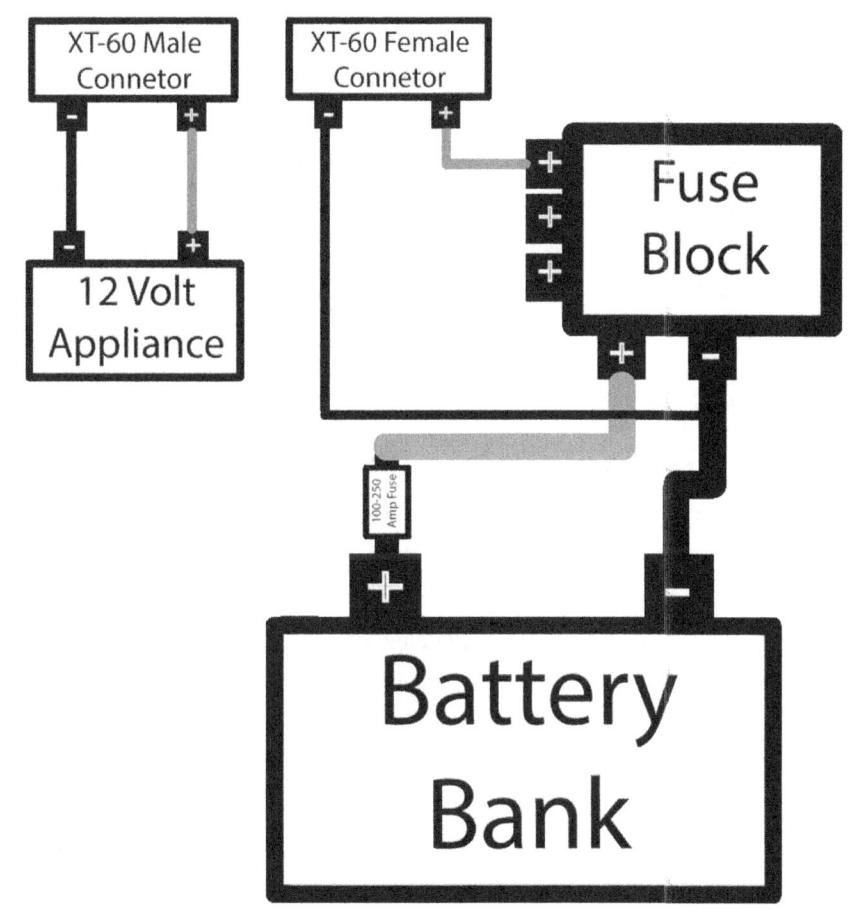

In order to use an XT-60 connector, you will need some soldering experience, or buy them with the wires attached, and crimp them instead.

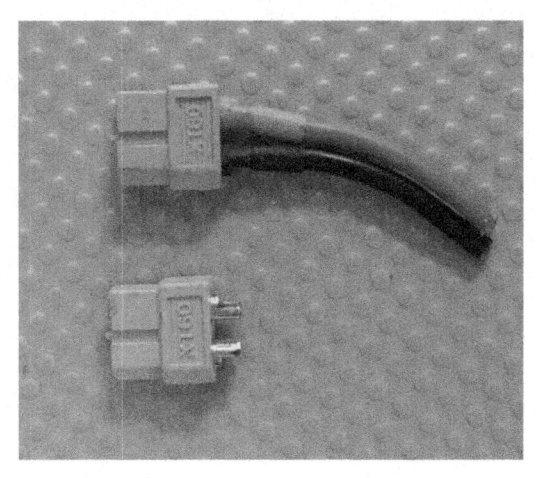

If you choose to solder it, watch an online video because it can be a little bit tricky.

Use a "helping hands" tool, or vise clamp to hold the connector.

1. Heat the terminal and add a little solder.
2. Choose a wire that you wish to attach to the connector. Tin the wire with a small drop of solder. Add heat shrink if you wish.
3. Next, heat up the connector terminal and insert the wire. Wait until the solder becomes really hot and shiny.
4. Remove the soldering iron and let it cool while the wire is supported inside of the terminal. Be sure not to overheat the connector because it can melt.

One thing that I dislike about xt-60 connectors is that you cannot use them with large wires. If you need a large wire gauge power connector, you will need a "Anderson Powerpole connector". These come in a range of sizes and work extremely well.

Powering a Laptop without an inverter

Powering a laptop with an inverter is a bit more inefficient than powering it with DC power, directly from your battery. If you use a laptop often, you will want to power it directly from a 12-volt source with a voltage converter. This will take your batteries 12 volts, and bump it up to 17/19/24 volts, or whatever your laptop requires.

The easiest way to do this is to find a laptop charger that can plug into a cigarette lighter. These chargers usually come with a variety of adapters so that they can work with different laptops. Usually, they can output different voltages to work with different laptops. If this works for your needs, you can splice the two power wires into your system, and attach it directly to the fuse block like any other appliance. You can also wire it directly to an XT-60 connector plug so that you can plug it in at different locations of your vehicle (if you have female plugs wired around your vehicle).

If you want to complicate things, you can buy what is called a bulk DC-DC converter. This simple device takes your solar system's 12-volt power and converts it to whatever voltage you please. This complicates things because you will need to find a plug adapter so you can connect the converter to the laptop.

Using a Bulk DC-DC converter can be incredibly useful for powering strange appliances that require weird voltages.

Adding efficient interior lights to your vehicle

Led's or "light emitting diodes" are what you need to light up your vehicle. They are 15% more efficient than fluorescent lights and 6 times as efficient as incandescent lights. In a solar system, this matters a lot!

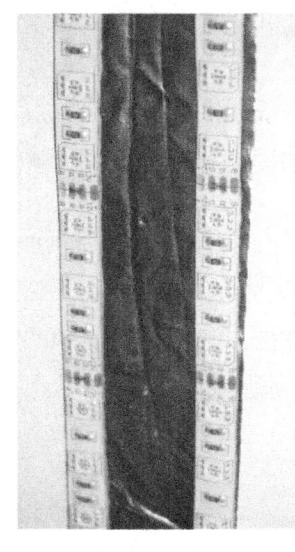

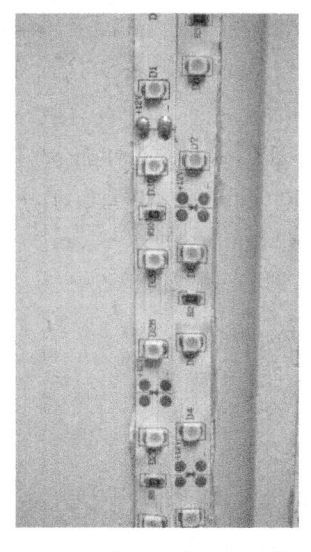

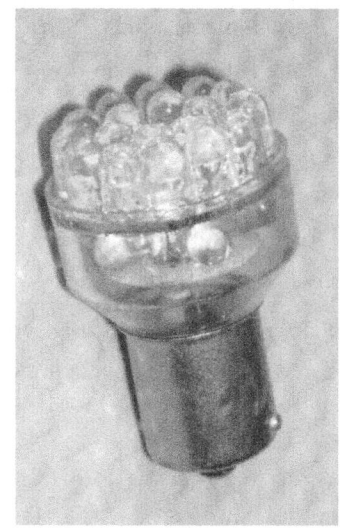

- Be sure to fuse your LED lights properly! At most they will consume 5 amps, so a 7 amp fuse would be perfect. Be sure that the wires supplying power to the LED controller are around 14 gauge (depending on the length the wires need to run). LED controllers love to develop short circuits from poor circuit board design and faulty soldering joints. Most LED controllers are made overseas and the quality of parts used is not the best, so a fuse is always required.
- If you have a vehicle with preexisting incandescent bulbs, buy a LED replacement bulb. They work well and take about five minutes to install. Inspect the solder joints and wiggle them to look for faulty connections.
- If you plan to live in your vehicle full time, try to use red green blue LEDs. These allow you to change the color to anything you please. The reason this is helpful is to switch the lights to a red hue so that they do not keep you awake. If you use white light LEDs, you will have a hard time falling asleep.
- Buy a nice flashlight. Do not use a cheap 5 dollar LED flashlight. Find a rugged, waterproof and rechargeable flashlight that is actually expensive. You will wonder how you lived without it for so long.
- You can wire up some waterproof floodlights to the exterior of your vehicle. These are particularly useful if you use them in combination with peepholes. If someone is trying to mess with you or your vehicle, you will either scare them or see what they are doing. I highly recommend this method!
- LED light strips are fun to use. You can mount them inside cabinets and corners. You can buy corner mounts that come with a light diffuser if you want your rig to look like a space station. The possibilities are endless and most solar systems can power a ton of LEDs quite easily.

Most LED strips come with connectors so that you do not have to solder them. If you are not skilled at soldering, stick to using the solder-free connectors to attach the strips together.

So fill your rig with light! Try to position the light switches in a centralized location, and try to mount the LEDs on something that will stay cool. Besides that, it's pretty straight forward. 2 power wires connected to your fuse block, and you are set!

Switches

This can be a basic set up consisting of individual toggle switches, or you can build a space shuttle dashboard with light up switches of every color. Really depends on how creative or lazy you are.

- Use a switch that is rated for whatever amperage you plan to use it with.
- Attach the switch to the positive wire of any appliance.
- Buying a board that comes with multiple switches is nice. This way you can supply the board with a nice thick power wire from the fuse block, and supply your lights and other appliances with the switchboard's power. Having all of your switches in one location is ideal.
- Add some accessory switches and power lines so that you can test new gadgets without splicing into your battery bank or fuse block.

If you are too lazy to wire switches to everything, you can use XT-60 connectors instead. Simply unplug your appliances when you are done using them.

Temperature Regulation Appliances

In most vehicles, temperature regulation can be difficult. Vehicles are typically made with materials that do not insulate well, which means that temperature fluctuations can be extreme. If you have a solar power system, you can put the electricity to work to control the temperature of your vehicle.

Efforts at vehicle temperature regulation will be futile if you do not take steps to insulate your rig and/or your own body.

Cooling your vehicle

- **Fans:** Particularly roof vent fans. If you don't have one, you will cook. If you live in an extremely hot area, you will need to also install a fan that sucks air in from the bottom of your vehicle. You may need to add a filter to prevent dirt and dust from entering your vehicle. If you have a floor fan that pushes air into the vehicle, and a roof vent fan that pushes air out at the top, you will be good to go!
-There are DC fans and AC fans. If you plan to run the fan all day, and you have a small system, you will need a DC fan. If you have a large system, then use your inverter to power a large AC box fan. Mount it in a window or face it directly at you.
-Fans are my favorite method for cooling, and my current RV has 4 large AC fans wired to Wifi smart plugs so that they turn on at certain times of day, or if it gets too hot. 3 fans in the roof, and 1 fan that sucks air into my RV. I also have a HEPA filter on my air intake fan so the air is cleaned before it enters.
- **Air Conditioner:** These work well, but require a lot of electricity. If you have a 1200+ watt solar system or a large generator, go for it. If not, I would avoid using them. If you absolutely need air conditioning powered from solar, use a 500 watt window air conditioner and a large pure sine wave inverter. Keep in mind that even if you have a lot of sunshine, it will be tough to power any air conditioner for a prolonged duration. Buy the largest battery bank that your vehicle can handle. You will need it.
- **Swamp Cooler:** Bad idea in my opinion. Unless you live in the desert, an air conditioner is a better idea.
- **Compressor fridge gel packs:** If you have a super-efficient 12-volt compressor fridge, buy some gel packs and leave them in the fridge. If you get too hot, use the gel packs to cool yourself off.
- **Paint your RV white:** Not as hard as people would think. RV fiberglass and exterior white house paint with primer work well! If you have an older rig, I highly recommend painting it white.
-I have seen black RV's and wonder how much energy it takes to cool the interior during summer. If you have a black RV, you better have a large generator, a few air conditioners and a large wallet for gas.
- **Plasti-dip your windows white and insulate them:** This has made the biggest difference for me. My RV has large tinted windows in the living area and I could not cool the RV down even with an air conditioner. What solved my problem was plasti-dipping the windows and window frames on the outside, and using foam insulation on the inside of the window. This made my RV feel like a home, and I cannot recommend it more!

Heating your vehicle

You can either find ways to heat up only your body, or you can heat up your entire vehicle. Heating an entire vehicle with electricity can be difficult. For most vehicles, using a small propane heater can be a game changer. One small canister of propane will offer more heating power than a giant battery bank. If you want to use only solar power to heat yourself up, you will need to focus on methods to insulate and heat up only your body. Buying an expensive down jacket and pants, used in combination with the methods below, can make mobile electrical heating feasible.

- **Seat heaters:** My favorite method. These wire directly to 12 volt systems and generate a lot of heat. You can attach them to xt-60 plug connectors, and hook them up only when needed. Mount and hardwire them into your bed. Wire them to a switch next to your bed. In order to make this system safe to use, you should hook up indicator lights. Wire some lights in parallel so that when you turn the seat heaters on, a light will come on. Mount this light in an easy to see location. If you leave the seat heaters on, and you leave your vehicle, they could melt, or worse, catch fire. If you use a cheap seat heater (usually available for 12-20 dollars online), you will have problems. These cheap seat heaters develop short circuits and like to overheat. You must buy the higher quality seat heaters (usually 75 dollars for a set). These can be used continuously, for years, and will keep you extremely warm.
- **Electric blankets:** If you have a Pure Sine Wave inverter, you can run electric blankets. A lot of people love to use these, but using an inverter to power a heating appliance is terribly inefficient. If you have a lot of power, and a large system, go for it. The 12-volt seat heaters above will still be better.
- **Infrared lamps:** You can wire these to most inverters, including the cheaper modified sine wave inverters. They may create a buzzing noise, but it's fine. You can point these lights directly at yourself, and they can heat you up quickly. They are also pretty cheap. The downside to these is that they require an inverter, which means that you will have some losses. Again, 12-volt seat heaters are better, unless you have a large system.
- **Radiant heaters:** Work well, but require a lot of electricity. If you have a generator, they work well.
- **Convection heaters:** I do not recommend these for vehicles unless the vehicle is well insulated. These heaters are made to heat up an entire room, and not an individual person. All things considered, even if I had a lot of electricity, I would use other methods to heat a vehicle.
- **Microwave rice bags and gel packs:** You can use your own microwave, or a gas station/ grocery store microwave. If you heat up a large gel pack or rice bag and go to sleep with it, it will keep you toasty warm.
- **Hot water bottle:** You can use a water bottle and fill it with hot water, and you are set! Going to sleep with one of these is awesome. But you need to make sure that it will not leak. Many pharmacy stores sell rubber hot water bottles made for this purpose. All you need is hot water, which you can do with a stove, or an induction stove top.
- **Large load appliances create heat, so use it:** Most appliances such as refrigerators, computers, amplifiers, inverters will produce excess heat. Mount these appliances in strategic locations so that the heat they create can warm your vehicle.

Other Methods

Passive heating and cooling can be done with a vehicle: If the sun is shining on one side of your vehicle, use it to heat up the vehicle. If you need to cool down your vehicle, bring in cool air from the shady side of the vehicle. The easiest way to control this effect is to modulate heat entering the windows with insulated window covers that can be added and removed quickly. You can take this a step further and cover entire walls with blankets. Vehicles and the sun move continuously, so you need to be able to change the insulation of your walls and windows without much effort. Velcro comes handy for this method.

Moving the vehicle to a colder or hotter climate: Vehicles are mobile. If it is too hot, drive to the ocean, towards the poles, or find some shade. If it is too cold, drive to the equator. Microclimates can also make a huge difference. Look at a weather map in your area and scout out locations with better weather.

Add more solar panels: This will insulate the roof of your vehicle from the sun and create more electricity. It is a win-win situation.

Emergency method: If it is very cold, you may need to make a small tent of blankets inside the vehicle, and add an indoor-safe propane heater. If your vehicle is large, this may be the only way to heat yourself up. Having some emergency blankets on hand is a good idea if you spend a lot of time in cold environments. Do yourself a favor and drive toward the equator if you are dealing with extremely cold temperatures.

Buy a smaller vehicle: If you have a large RV, you will require an enormous amount of energy to heat and cool it. If you have a small van or RV, you will always be able to modulate the temperature easily.

How to use a Bulk DC-DC Converter

If your laptop, or appliance, requires a strange voltage, and you cannot find another way to power it, you may need to use a bulk DC-DC converter. These converters will transform your systems 12 volts to any voltage you want.

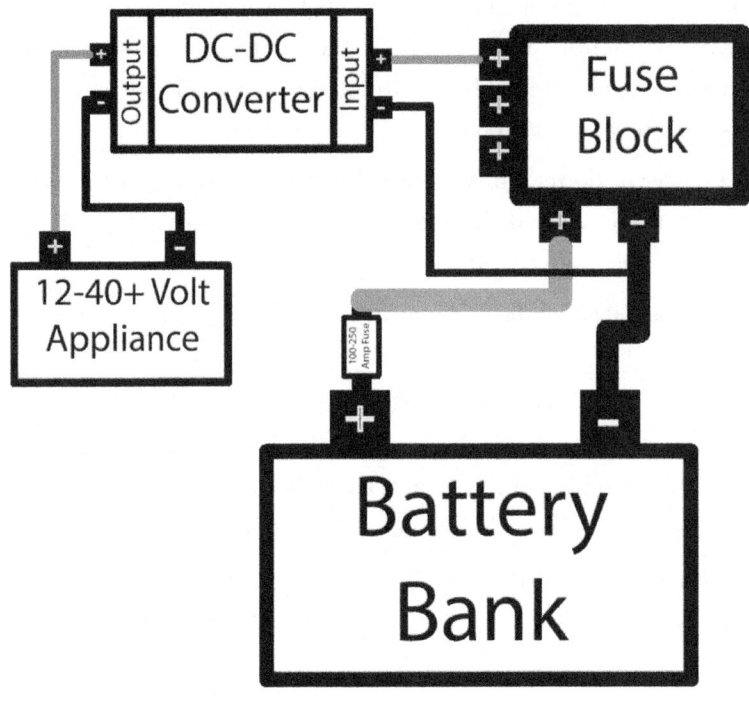

1. Connect the converter to your system with 2 wires, a positive and a negative. Attach these wires to the "input" terminals of the converter. When you power up the converter, be sure to not touch the output terminals. These terminals are usually connected to a small electrical component that can cause a shock. So be careful and avoid touching the output terminals.
2. Next, attach a voltmeter to the output terminals. Check the voltage. Most converters will have a small screw that you can turn, on the converter itself that will allow you to change the voltage output. Turn this screw and watch your volt meter. If the voltage changes while you turn the screw, you have found the right one. Turn the screw until the voltage output is what your appliance desires.
3. Disconnect the power converter from your system, and use a metal tool to short the output terminals. This will discharge all of the capacitors that could potentially shock you.
4. Now you can attach your appliance. Run wires from the output terminals of the converter to your appliance. Be sure to connect the positive output terminal to the positive appliance wire. If you install your appliance with reversed polarity (mixing up positive and negative wires), you will destroy the appliance.
5. Now the moment of truth! Power up the converter and see if your appliance works.

Keep in mind that this can be dangerous. If any of the wires are wrong, or you supply the wrong voltage, you will destroy the appliance.

Adding AC Appliances

Buy a few extension cords and power strips, and plug everything in. Things to keep in mind:

- Buy the proper gauge AC extension cord for your appliance. If you plan to power a microwave with your inverter, you must buy an extension cord that is rated to carry the wattage of the microwave.
- Induction loads, such as motors and microwaves, require an increased amount of power to start up. If your inverter is not rated to handle the initial amount of required power, it may be unable to power it. As an example, let's say you buy a 1000 watt microwave and a 1000 watt inverter. There is a good chance that this combination will fail to work. This is because the microwave will probably use 2000 watts just to start up. This will shut down the inverter, and you will be confused. If you want to power a 1000 watt induction load, you will probably need a 2000 watt rated inverter.
- If you have a large inverter and you are having a difficult time powering large loads, there is a good chance that you need thicker inverter cables. Pushing 200+ amps of 12-volt power is difficult for any wire that is less than 0 gauge in size. If you need to power a large induction load, you will need inverter cables that can handle the startup power requirement. Always buy inverter cables that are thicker than what is necessary. Also, feel the cables with your hands while the inverter is powering a large appliance. If the cables are becoming really warm, it is a sign that you need a larger cable.

Off Grid Internet

There are a few ways to go about doing this:

- Tethering a cellphone internet connection with Wi-Fi or a USB cable (requires an smart phone app)
- Hotspot service provided by a cell carrier
- Using a large antenna and powerful Wi-Fi adapter to connect to free Wi-Fi spots

But these options pale in comparison to a new set products that have recently entered the market:

4G LTE Router with High Gain Antenna

This device will allow you to have fast internet on-the-go. There are various 4G routers on the market, so here are some features that you need to look for:

- **High-speed "AC" Wifi for Multiple Devices:** If you have multiple smart devices, or smart appliances, this is a must.
- **Ethernet Cable Input:** This will hardwire your computer to the router which will enable faster speeds than achieved with Wifi. Required if you plan to stream 4K or play online video games.
- **4G LTE Directional Antenna Upgrade Options:** Not required, but very useful if you plan to use your router in rural locations. After you upgrade your antennas, point them in the direction of a cell tower for a boost in signal. To find out where a nearby cell tower is, download a "cell tower finder" app for your phone.
- **Low power consumption AC adapter/ 12 volt input/ USB input:** Most 4G routers come with a small AC adapter that powers the router. These adapters do not require much electricity, but they do require you to run your inverter 24/7, which can require a lot of electricity. If you have a large system and lots of sunshine, running your inverter 24/7 is very easy to do. If you have a smaller system, you may need to buy a router that can be powered directly from your battery. Better yet, there are some routers that can be powered with a USB! These are great if power generation is limited.
- **Mounting Options:** My current 4G router comes with 4 screws so I can mount the router where I please. Not all routers have this option. If location of your router matters (for heat dissipation or signal boosting reasons), look for a router with small holes so that you can screw-mount it anywhere.

These units are easy to setup and use, but they do require a SIM card from a cell carrier. Any cell carrier will work, but the SIM card needs to be specifically enabled for hotspot service. If you stick your cellphone's SIM card into your router, it will probably not work. Buying a hotspot SIM card usually has a monthly cost, but it is worth every penny! Having unlimited internet 24/7, anywhere I go is game changing. If you want to see what router I am currently using or my SIM card service, check out my website: http://www.mobile-solarpower.com

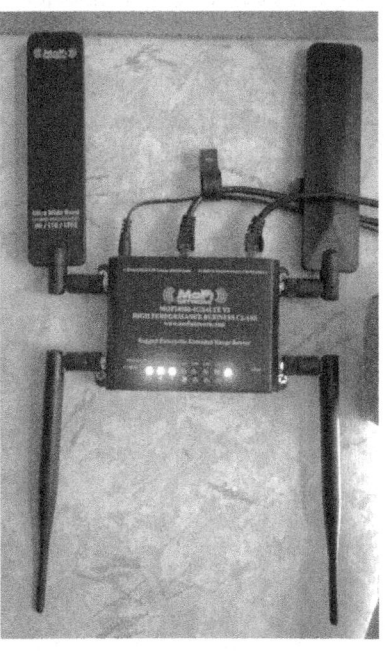

Smart Home Appliances

If you have a 4G LTE router in your vehicle and a way to power it 24/7, you will be able to use smart home appliances! These are appliances that can be controlled anywhere in the world with your smart phone or a computer, or can be voice-activated when you are present. You can also run them on schedules so that they control themselves. Here are my favorite smart home appliances for off grid vehicles:

AC fans and a Wi-Fi controlled Smart Plug: By combining these two devices, you can turn your vehicles cooling fans on at any time, from anywhere, with your smartphone. The smart plug can also turn the fans on at certain times of day, or when the weather for your area reaches a certain temperature. If you have pets, this is a must!

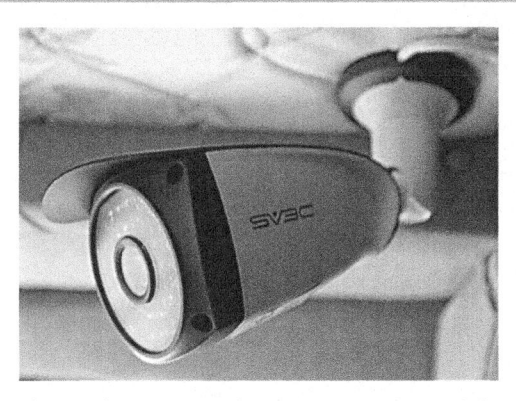

Wifi connected or Ethernet wired Security Cameras: This will allow you to see inside your vehicle when you are not present. You can also have the security camera send an alert to your phone if it detects movement or a loud sound. The most important features are:

- **Wide angle lens:** Do not buy a security camera that requires you to control the cameras movement. A wide angle camera is much easier to use.
- **Night Vision:** Typically security cameras have night vision, but some come with infrared LED lights which increase the night vision capabilities drastically. The more LED's the better!
- **Motion Detection and Phone Alerts:** Not all cameras have this, and it is extremely useful.

- **Ethernet Cable Connected Cameras are better:** No matter how much money you spend on a security camera, a Wi-Fi connected camera will not work nearly as good as a hardwired Ethernet cable camera. Wi-Fi cameras, in my experience, intermittently fail to connect or have lag. A hardwired camera connected directly to your router will ensure a solid connection at all times.
- **Cloud video recording storage is recommended:** Usually this costs about 5-10 dollars a month, but it is worth every penny.

Wi-Fi Home Assistant: I use a google home mini so that I can control my smart appliances with voice commands. I also use the assistant for answers to science or history questions, or to look up a word definition. These are not for everyone, and some people do not like large tech firms listening to their every word, but I love it!

LED Lights Controlled with a Wifi Smart Plug: This combination will allow you to schedule or remotely operate your lights with your smart phone. I love this combination because if I know that I will not be home all night, I can remotely turn on the lights to deter thieves. You can also set schedules for the lights to turn on or off at different times of day. You can also connect the smart plug to a home assistant smart device so that you can use voice commands to turn on the lights.

Solar System Maintenance Schedule

Every day:

- Check the voltage of the battery bank

Every month:

- Visually inspect the batteries for damage, leaks, corrosion and faulty battery terminal connections (wiggle the wires).
- Inspect all wires and fuses. Look for damaged wire insulation caused by vehicle vibrations, animals, chemicals etc. Wiggle the wire connections at the fuse block and inverter to ensure a strong connection.
- Inspect and clean the solar panels. Also look for cracks, damaged wire insulation and tug on the panels a little to make sure that they are securely attached.

Every six months to a year:

- Remove and inspect battery terminals and inverter cables and clean them with a wire brush. Be sure to disconnect the solar charge controller from the solar panels before you remove any battery cables.
- Put a clamp amp meter on individual solar panel wires to measure the output of each solar panel. If one solar panel is not producing as much power as the other panels, a short may have developed, and the panel may need to be replaced.
- Inspect solar charge controller terminals. Unscrew each terminal, take the wire out, inspect the wire strands, and reconnect it. Sometimes the charge controller terminal will damage the wire, and you will need to re-strip the wire and reinsert it.

Besides that, use your system as much as you want, and enjoy the free electricity! Most solar systems should not require much maintenance at all besides cleaning the solar panels from time to time.

If you have vented batteries, you should follow the maintenance recommendations provided by the manufacturer.

If your battery bank has a lower voltage than usual, or it seems to have a reduced capacity, you may have a short in your battery, or the batteries are too old or may be damaged. Remove the questionable batteries from your system and have them inspected by an auto parts store (usually you can have them load tested and checked for free). Over time, the only thing that you need to replace in your system will be the batteries.

Over time, the solar panels will produce less electricity, but they should still produce electricity well after you die.

Odds and Ends

Efficient Computer Options

The easy way to do this is to buy a laptop, tablet or a powerful phone. If you plan to run games or video editing software smoothly, you will probably need a desktop computer. I am currently running an expensive gaming computer and large monitor in my RV and it uses 300 watts, which is fine because I have 1000 watts of solar on my roof. Most mobile systems would have a tough time powering such a large computer setup.

Luckily, there are some efficient desktop options that use the same power as a laptop. Typically, the smaller the housing of the computer, the less power it will use. If you need to run strenuous programs, you can buy a "small form factor" gaming desktop computer.

Regardless of the type of computer that you plan to use, I recommend using a solid-state hard drive and the latest chipset available. Once you run this combination, you will never be able to use a slower computer again. It will spoil you to no end.

Phantom Loads

If your system is not performing as expected, and you are sure that the solar system is working, and the battery bank is healthy, you may have a phantom load. This means that an appliance, or possibly a small short circuit, is leeching power from your system. If you have a small system, and an appliance is leeching 40 watts continuously all day, you will have problems! Usually, these ghost loads will surface during the winter when solar power production is at its lowest.

How to find a phantom load

- Disconnect all solar power from your battery. This can be done by removing a main solar panel array wire at the solar charge controller.
- Turn off all appliances.
- Use a shunt, or amp meter of some kind, to determine the amp usage at the battery, or at the fuse block for individual appliances. This can be done by removing a fuse and inserting an amp meter, or by using a clamp amp meter. If you do not have an amp meter, you may be able to rent one. Amp meters typically have a limit to how many amps they can measure before they are damaged. If you are using a multimeter as an amp meter, the typical limit is 10-20 amps.
- If your amp meter shows a current draw larger than .1 amps, you need to look around your vehicle and find out what is using power. Having a fuse block proves useful in this situation. Remove individual fuses, and check each appliance for an amp draw. Some appliances will use a small amount of power whether it's on or not. An inverter uses 1-3 amps if it is turned on, regardless of if anything is connected to it.
- If an appliance uses too much power when it is turned off, you may need to wire a switch to the appliance, so you can manually turn it off when it is not being used.
- If you look everywhere, and cannot find the source of the phantom load, you may need to run your usual appliances, and see if anything is consuming a lot of power. Also feel the wires to see if any of them are

getting warm. This could be a sign that an appliance has developed a short circuit, or that the wrong size fuse and wire is being used.

Sometimes a phantom load can drive you crazy! If your wires are not organized, it can take hours to find them. If you have a fuse block, and each fuse is wired to only one appliance, you should be able to find the phantom load quickly.

Storing a Solar Power System

If a vehicle with a solar power system is stored indoors, your batteries will discharge themselves over the course of a few months to a year. This is fixed by using a float charger. These are available online for a cheap price and plug into a wall outlet.

If you plan to park the vehicle outside in the sun, you should be good to go. For extra safety, disconnect the fuse block from the battery bank.

Connecting Different Types of Solar Panels Together

This is not recommended, but over time you may collect various sizes and types of solar panels and may need to wire them together. It is safe to do but you will typically experience some energy loss. There are two golden rules that you must follow:

1. If a group of solar panels have the <u>same voltage rating</u>, connect them together in a <u>Parallel</u> configuration.
2. If a group of solar panels have the <u>same amp rating</u>, connect them together in a <u>Series</u> configuration.

As long as you follow those two rules, you can connect any size of solar panel together. Even if you connect a 1000 watt solar panel to a 1 watt solar panel, if you follow the two rules above, it should work fine. Here are some other tips to follow:

- If the solar panels you wish to connect are different voltages and different amp ratings, use a parallel connection. Do not do this if the voltage difference is extreme! Do not try to hook a 1 volt panel to a 30 volt panel.
- If a solar panel does not have diode protection, do not connect it to other solar panels.
- If you have a really old solar panel and you want to connect it to an array that has brand new solar panels, use a parallel connection (unless the volt rating difference is extreme. Check it with a volt meter).
- Do not series connect solar panels that are not of the same age or wire gauge.

After you connect a group of different solar panels together, check their voltage. If they are producing electricity, you should be able to connect it to a solar charge controller.

Connecting Different Solar Charge Controllers to One Battery Bank

This should not cause any problems, and is safe to do. Just make sure that all of the solar charge controllers are designed to charge the type of battery you are using.

I have connected 4 solar charge controllers to a battery bank and never had a single issue. I also did this for years, and had 2 MPPT charge controllers, and 2 PWM charge controllers. I am currently running 2x large MPPT controllers to a battery bank, and it works great!

Solar Electric Cooking and Food Preparation

Common Off-grid Cooking Appliances:

- Induction stove top
- Microwave (no they are not bad for your health)
- Blender
- Electric Pressure cooker (or traditional pressure cooker used on an induction stove top)

You can power anything you please with a large enough inverter. Even if you have a small system, most large load appliances can still be powered safely for a short time. Using a 1000 watt blender for 30 seconds does not require many watt hours at all. If you plan to use any of these appliances for a prolonged duration, such as an induction stove top to cook a soup for an hour, you may have a problem.

A solar oven is the most efficient way to use the sun to cook food. Some solar ovens work so well that you can cook food on cloudy days. They can be built with aluminum foil and cardboard. Online search "DIY Solar Oven" for ideas.

Solar Water Heating

Instead of using electricity or gas to heat water, use the sun! Solar water heating is a great way to take a hot shower for free!

These systems can be very complex, or very simple. Here are two basic ways to heat water with the sun:

- Create a shallow wooden box that has water pipes going back and forth inside. Paint the pipes black, and direct sunshine at the pipes. Mount this box on the roof of a vehicle. Use a water pump to run water up to the box and have a return pipe bring the heated water to a shower or hose.
- Or use a black outdoor shower bag, fill it with water and leave it in the sun for a while.

The outdoor shower bag method is ideal in a vehicle. When you are not using it, fold it up and pack it away somewhere. They can even be mounted on a roof, but it will get pretty dirty over time.

Should you install a battery isolator?

A battery isolator connects the charging system of your vehicle to your battery bank. This causes your battery bank to charge while you drive (it only works when you drive your vehicle. When you turn your vehicle off, it disconnects the battery bank from the charge circuit). This prevents the possibility of draining your vehicle's starting battery.

Over the years I have learned that the charging system of most vehicles is not designed to charge a deeply discharged solar battery. It can cause premature failure of components, like the alternator. I used to recommend that people install a battery isolator for smaller systems, but I have since changed my mind due to the problems that it can cause in the long run. It is best to keep the two systems completely separate. You can install a large wire and switch between the starting battery and your solar battery bank for emergency situations, but carrying some jumper cables would be much easier to do.

Increasing solar output by reflecting light onto your solar panels

A really fun way to increase the output of your current solar system is to build large reflectors to direct more sunshine onto your panels. I have experimented with this, and it did increase the output of my panels, but it was not a sustainable effort. The reflectors needed to be taken down every time I wanted to drive, or whenever it was windy. The reflector I built was simply a large reflective square made of PVC pipe with an emergency blanket covering the entire thing. I mounted the reflector on the side of my RV and would raise it up if I wanted to use it. The most difficult part was finding the best angle so that it would shine light directly at the solar panels. The sun moves quite fast when you need a good angle, and I found that I had to change the angle every couple of hours to keep up with the sun.

So it's a fun project idea, but for most vehicles, not feasible. If you have a stationary system, then go for it!

Was this book useful? Let me know by leaving a book review on Amazon.com ☺

Be sure to check out my solar power website!
http://www.mobile-solarpower.com

Thank you for reading! I hope you love your new solar power system ☺

Notes

Made in the USA
Monee, IL
26 May 2022